Beauty - Cosmetic Science, Cultural Issues and Creative Developments

Edited by Martha Peaslee Levine
and Júlia Scherer Santos

Published in London, United Kingdom

IntechOpen

Supporting open minds since 2005

Beauty - Cosmetic Science, Cultural Issues and Creative Developments
http://dx.doi.org/10.5772/intechopen.77628
Edited by Martha Peaslee Levine and Júlia Scherer Santos

Contributors
Juliana De Souza Alencar Falcão, Valdicléia Massilon De Abreu, Maria Da Glória Batista De Azevedo, Jumina, Harizal, Yehezkiel Steven Kurniawan, Vladimir Testov, Daniel Gill, Pascale Weber, Li Yun, Rong Rong, Júlia Scherer Santos, Levan Torosian, Wenting Mu, Fan Wu, Mavis Lilian Henriques, Debasis Patnaik, Poonam Prakash, Rahul Bahri, S.K. Bhandari, Martha Peaslee Peaslee Levine

First published in London, United Kingdom, 2021 by IntechOpen
IntechOpen is the global imprint of INTECHOPEN LIMITED, registered in England and Wales, registration number: 11086078, 5 Princes Gate Court, London, SW7 2QJ, United Kingdom
Printed in Croatia

British Library Cataloguing-in-Publication Data
A catalogue record for this book is available from the British Library

Additional hard and PDF copies can be obtained from orders@intechopen.com

Beauty - Cosmetic Science, Cultural Issues and Creative Developments
Edited by Martha Peaslee Levine and Júlia Scherer Santos
p. cm.
Print ISBN 978-1-83962-447-6
Online ISBN 978-1-83962-448-3
eBook (PDF) ISBN 978-1-83962-449-0

Meet the editors

Dr. Martha Peaslee Levine is a psychiatrist who specializes in treating eating disorders and works with young adults as they tackle life's challenges. She is Medical Director of the Eating Disorders Partial Hospitalization and Intensive Outpatient Programs at Penn State Health. In addition, she is Interim Director of the Office of Professional Mental Health at Penn State College of Medicine. Dr. Levine has spoken both nationally and internationally on many issues related to eating disorders. She has lectured and led workshops on the benefit of journaling and authored a blog, Your Write to Health, on Psychology Today. Dr. Levine has published several children's books and believes that we all need to discover beauty in the world around us and in ourselves.

Júlia Scherer Santos is a pharmacist with a Ph.D. in Pharmaceutical Nanotechnology from the Federal University of Rio Grande do Sul, Brazil. She is a specialist in cosmetology, cosmeceuticals, and aesthetic health. She works in the areas of cosmetology, nanotechnology, pharmaceutical technology, health, beauty, and aesthetics. Dr. Santos also has experience as a professor of graduate courses.

Contents

Preface

While serving as the editor for the book The Perception of Beauty, I was struck by all the different iterations of beauty. Through my work in the field of eating disorders and at Penn State College of Medicine, I have found that many young adults struggle with poor body image and low self-esteem. We are all inundated by media images that push us towards a vision of perfection. The video Evolution by Dove shows how many alterations go into creating media images. There is a staff for hair and makeup, which take hours to do, and then the images themselves are altered digitally to make them even more "beautiful." This impossible persona is then presented to the public as an ideal, yet it is not attainable in real life.

When I edited that first book, I expected stories about body image. How could that not be a focus with all the overwhelming messages related to beauty that are perpetuated by the media? Some chapters did address those issues, but others looked at different views of beauty in art and science. I learned how beauty is perceived and presented in a variety of venues. That same pleasant surprise came with this volume as well. While some of the chapters do focus on eating disorders and body image, others take us down additional avenues so that we can consider the beauty of creative thought and scientific alterations in the pursuit of beauty.

Beauty - Cosmetic Science, Cultural Issues and Creative Developments is divided into three sections.

Section 1 encourages us to recognize beauty in nature and ourselves.

Chapter 1, "Sense of Beauty and Creativity," considers how beauty leads to contemplation, observation and, ultimately, to creativity. When we witness the world around us, we discover new connections to our thoughts, to each other, and to our limitless possibilities.

Chapter 2, "Environmental Art and Wilderness: The Stakes of Beauty," allows us to travel with the artist as she explores places of wilderness. In the words of the author, "...it is a question of thinking of the place from the perspective of the body and conversely of thinking of the body from the perspective of the place in which it is immersed." Through this work, readers are invited to experience how the wildness of nature can stir the same wildness in our souls.

Chapter 3, "Identity, Self-Identity and Beauty in Chinese Female Worker," provides a beautiful and rich dissection of the poem "The Sundress," written by Wuxia, a migrant worker in Shenzhen. The author leads us through the poem's imagery and interweaves an understanding of the role that culture can take in shaping one's sense of beauty and self-acceptance.

Section 2 looks at formal and natural sciences as they related to Beauty.

Chapter 4, "Beauty in Mathematics: Symmetry and Fractality," examines opposite ends of beauty—from stable order to the self-organization of chaos. While we can all readily accept beauty in symmetry, it may be harder to see beauty in chaos. Yet, the author encourages us to see the beauty of both sides.

Chapter 5, "Cosmetovigilance in Hair Straighteners: Determination of Formaldehyde Content by Spectrophotometry and Label Evaluation," examines the formaldehyde content in various hair straighteners. Society often pushes individuals to use products to obtain a "standardized" beauty that can put them at risk. The results lead to a call for vigilance and oversight to ensure safety for consumers.

Chapter 6, "Nanocosmetics: Production, Characterization, and Performance Improvement," looks at nanoparticles that are used in certain products, such as sunscreens and moisturizers. These particles have improved ultraviolet protection and help with the controlled release of active ingredients. This chapter provides an understanding of products designed to help protect our skin and its natural beauty.

Chapter 7, "Chalcones in Dermatology," examines the role of chalcones to treat a number of skin conditions, such as vitiligo and psoriasis. Our skin is the canvas of our external beauty. Many individuals struggle with skin conditions that can then also affect their sense of self. Our external beauty is often tied to our self-esteem.

Chapter 8, "Maxillofacial Defects: Impact on Psychology and Esthetics," examines beauty as it relates to alterations of the face. The authors examine the effects facial defects can have on the psyche of the individual and in interpersonal relationships. They offer guidelines to help clinicians consider the emotional toll of both the injuries and the rehabilitative procedures.

Section 3 examines interpersonal influences on external and internal beauty.

Chapter 9, "Decoding the Signals of Facial Attractiveness: A Communication Theory Perspective," looks at ways in which the face transmits a wide range of social signals. Using the communication theory perspective, the author discusses ways that faces and attractiveness provide instant tools that alert us to the potential for cooperation or aversion within social interaction.

Chapter 10, "Social Media and Its Effects on Beauty," looks at the impact that social networking sites have on the perception of beauty. The author looks at the positives and negatives of a "selfie culture," where individuals turn to social media sites for a sense of self-acceptance.

Chapter 11, "Blossoming for Whom? Social Approval and Body Image," delves further into the role of media and the approval of others as we evaluate our body image and sense of self. The author describes, "When we put a category of qualities on the pedestal, we are essentially labeling people who do not possess such qualities as inadequate." How do we maintain a positive view of ourselves when we are subjected to messages that diminish certain types of beauty?

Chapter 12, "Do Individuals with Eating Disorders See Their Own External and/or Internal Beauty?" considers some of the reasons individuals with eating disorders struggle to appreciate their own beauty, internally as well as externally. It encourages us all to become more compassionate to ourselves and to strive for self-acceptance rather than rigorous perfectionism and self-hatred.

While I was beginning work on this book, our family suffered the loss of a friend, Anna, who was a beautiful person, both inside and out. She pulled everyone she met into her orbit of positivity. She was full of energy and interested in everyone and everything, from public transportation to chocolate. Her memory reminds me of the goals of this book—to find beauty in our lives. To find beauty in all of its glorious forms. That is one of the joys in editing these chapters—people have many different views and insights into beauty. I hope that by reading this volume, you find a new perception, jog a new thought, or make a new connection to fuel your ingenuity. I hope that you work to practice self-compassion and find that beauty is there in both the symmetry and chaos of life. I hope that you recognize beauty in yourself and all those around you.

Martha Peaslee Levine, MD
Associate Professor of Pediatrics, Psychiatry and Humanities,
Penn State College of Medicine,
Hershey, Pennsylvania, USA

Júlia Scherer Santos, Ph.D.
Universidade Federal de Juiz de Fora,
Brazil

Section 1

Recognizing Beauty in Nature and Ourselves

Chapter 1

Sense of Beauty and Creativity

Levan Torosian

Abstract

This chapter discusses the impact of natural beauty on the human consciousness and the ability to be creative. The stages of formation and the sequence of the abilities that allow for the emergence of a creative life path and the development of inventiveness are described. A vision of how the sense of beauty develops in the childhood period, as well as how this feeling affects the ability to be inventive in various ways, is discussed. The relationship between the sense of beauty and such human properties as contemplation, observation, and ingenuity is defined. An analysis is made of what impacts our ability to see beauty. Some ideas about how structural changes in technology influence a person's technical environment and his ability to see natural beauty are provided as well.

Keywords: nature, beauty, creativity, contemplation, observation, ingenuity, technology

1. Introduction

The sense of beauty, its development and improvement, depends on many factors. Modern human society is trying to make people's lives more beautiful, comfortable, and technological. The artificial beauty created by some people is presented in various forms and offered for understanding and acceptance by other people. There is the sphere in which a person is invited to understand the beauty and somehow transform this understanding into the environment of his life, in choosing those forms of beauty that will bring him the greatest joy and well-being. A modern person needs to develop a sense of beauty in himself. Professional activity often requires ingenuity which is connected with the sense of beauty also. The influence of the sense of beauty on ingenuity, on the ability to create beautiful things with one's intellect, is undeniable. Therefore, let us try to understand what factors determine the development of a sense of natural and artificial beauty and what can become the basis for a harmonious transition from the perception of the beautiful to the creation of the beautiful. Readers are invited to become familiar with the analysis of such human properties as contemplation and observation. The relationship of the factors that create a transition from the process of perception of the beautiful to the creative principle and ingenuity is also considered.

When I hear the word beauty I ask myself—what is it? Is this a certain feeling or is it a way of thinking or is it an algorithm for perceiving phenomena? Does beauty make a person move along a creative path? Are all people equally able to perceive different beauties? Does the sense of beauty affect the development of human ingenuity in the technical field? We will try to consider these issues in this chapter.

2. The sense of beauty and nature

When we raise our heads to the sky and see how a plane gracefully flies at a high altitude, it seems to us beautiful, majestic, and ingenious, especially from the point of view of what we can create with the human's mind and hands.

It happened in my life that I was lucky enough to be born and grew up in a very beautiful place. The nature was connected to mountains, valleys, the sea, and good and beautiful people. From my birth I have been surrounded by natural beauty and harmony of human life. This fact predetermined all my further perceptions of life and movement along it, as well as my attitude to beauty.

I became a technical specialist after graduation, and at a certain point of my life, my feelings of natural beauty and creativity began to combine inside me, including in the field of technical ingenuity.

There are many different opinions and scientific approaches to developing a person's ability to be creative and to invent.

Creativity as the ingenuity is given to any of us, but many lose it as adults. Further development of these abilities depends mainly on the atmosphere in which the person was raised. I think that nature has the primacy role in this environment. It is the natural harmony of power and energy that determine for a child the very sense of beauty that will accompany him throughout his life. This feeling will also allow a person the ability to create something beautiful, regardless of what people do in a professional sphere.

It is very important to observe the diversity of natural wealth in childhood such as mountain topography, sunset and sunrise, the elements of rain and thunderstorms, sea storms and calm, and so on. Observing such phenomena, the child learns to feel and analyze, which is an essential necessity for the further development of inventiveness [1]. Natural playscapes offer sensory stimulation and physical diversity which is critical for childhood experiences [2].

It is in natural phenomena that the foundations of human behavior and abilities are developed, and in childhood a feeling of enthusiasm is formed. However, at this age, this feeling is still specific, not completely formed, and differs from the enthusiasm of adults, who, as we know can be admired in different ways. For example, a 3-year-old child is more likely to enjoy some toy than a beautiful sunset. This happens because in childhood, the brain of the child still accumulates information and pictures of the world in order, to build subsequently on this basis, its own unique algorithm of perception and behavior in the world around it. Therefore, those pictures of the world that a child observed in the first 5–7 years of his life determine the development of human abilities, the level of his susceptibility. That is why we observe such a variety of degrees and levels of perception, the ability to see and create beauty around ourselves in adulthood among different people and nations [3, 4].

The level of the perfection and development of our life has reached today is interrelated from nature. People continue to learn from nature. We copy natural phenomena, creations, and even in such a form as "ourselves." Natural beauty is in everything—these are colors and shapes; these are sequences of changing seasons, weather phenomena; these are life cycles; and so on. Thanks to the vision of this beauty, the human's brain is transformed through the process of observation, discovery, and development of new areas of science [5].

Technical devices that people created have allowed us to see a grandiose beauty that has been hidden until recently from humanity in the depths of space—the beauty of the universe and its breath [6]. So how wonderful it is that the perception of beauty on our own planet has expanded the scope of thinking and expanded the possibilities of seeing extraterrestrial beauty.

3. From the beauty to ingenuity

The study of human nature and world is associated with sequences and algorithms. From primitive actions to the highest complexity of engineering structures, a person in his thinking and behavior is based on algorithms and therefore on sequences. This ability has made it possible to create mathematics, physics, and other sciences that have become the basis of our prosperity. In all developmental processes, there is the beauty of discoveries, research, definitions, and relationships.

The development of a person's ingenuity is invariably associated with his observation, and the process of developing ingenuity can be presented as the following scheme:

Beauty → Contemplation → Observation → Creativity

From generation to generation, people pass on to each other knowledge, as perhaps the most precious legacy. The stratification of observations in different epochs during the existence of humanity has allowed us today to create very complex technical systems that surround us everywhere and give us a certain degree of freedom. Remember how many millennia people have watched the flight of birds, dreaming to rise in the sky and fly as well. Today, an airplane flying overhead seems to us to be the norm. However, in order to understand how to take off the earth and fly, humanity had to observe, observe, and observe.

How does the desire for observation appear? By the author's opinion, the natural beauty creates and develops in us this gift and ability to see the essence of beauty, since natural beauty was the first and natural call for a person to embark on a developmental path. When we meet natural phenomena or new places on our way where we are going, we sometimes cannot take our eyes off the beauty we see, and we want to look at it repeatedly. People do not get tired of natural beauty. Therefore, observation can be defined as the ability to learn from the beauties of nature.

In some discussions about the character of a person, his nature, one can trace the opinion that contemplation is a negative property. In some dictionaries, contemplative person is defined as a passive observer. Has someone ever considered how contemplation and beauty are connected? After all, if you analyze what a person is looking at more, then this is something beautiful, especially at a young age. So what is wrong with a person enjoying vision of beauty? Could it be the other way around? Those people who spend their life energy and time searching for something vital and do not have time to see the beauty of the world around them, their own beauty—maybe these people lose much more than contemplators do.

Of course, it is not impossible to be a one-sided person in the modern world, but a reasonable balance between the ability to contemplate and the ability to succeed in a profession can give a person more joy in life.

Creativity, as a way of self-expression of a person, is a companion for human society probably from the moment when people realized themselves. Nature has shown people the algorithms for the development of creativity. For example, there are algorithms of movements, algorithms of temperature and shape changes, algorithms of fire appearance, and so on. However, how could one know these algorithms without contemplation and observation? If we consider that contemplation is the ability to see for a long time and observation is the ability to see deeply, then these two properties become absolutely necessary for the development of inventiveness, as a process of manifestation of the abilities to create.

The invention of the wheel, which allowed humanity to race along the path of technological progress, was probably interrelated in nature from observations of how stones or logs roll, and natural fires made it possible for people to feel the heat and the benefits of using it on an ongoing basis.

Observations of nature gave a person an understanding of how to use forms in mathematics, architecture and engineering [7], and various colors in almost all areas of production and life. The beauty of natural sounds is transformed by a man in musical instruments. Finally, we observe the endless desire of engineers to create beautiful designs.

At the beginning of engineering, ingenuity lies in functionality, but soon engineers also focused on the appearance of their creations. Of course, it often happens that the beauty of structures is determined immediately during their design, but in the long term, the substantial part of the appearance always strives for new beauties, those that can only be born in the human's mind.

So why do not all people perceive the same phenomena equally and adults do not always associate beauty with the possibility of its subsequent transformation into ingenuity or creativity? Perhaps the reason lies in the ability to perceive beauty as a system of values.

If we look at what objects and natural phenomena surround us, then they will seem beautiful to us. Almost everyone agrees that we like beautiful landscapes, buildings, cars, ships, clothes, dishes, meal, paintings, music, and graceful movement. We love flowers, fruits of plants, natural stones, and in general, everything that can cause a feeling of delight. However, not everyone knows how to admire.

A person of a creative warehouse is able to see and hear beauty in almost everything as a rule. Persons, who do not have creativity, need additional environmental factors and perceptions of the world to delight them. Therefore, such people subconsciously strive to approach creative people to adopt the ability to admire from them [8]. In these cases, an additional responsibility lies with creative people—not to reject those who do not know how to create and be inventive and not to blame them for this but rather try to involve and teach them how to see true beauty, to be beautiful and enjoy beauty in life.

For many years, while working with students, I have observed how their worldview changes when they meet not only inventive teachers but also inventive students. The exchange of technical and cultural knowledge and skills during learning is the second stage in the formation of the ability to move from ingenuity to the perception of beauty, after the childhood period of life. It was during this period that the algorithms finally formed and what can make it possible to create something new on their basis.

When I was at school, my physics teacher read his poems to us in the classroom. Sometimes I wondered why he was doing this. However, when I began teaching students, I realized that many people could understand things that are difficult only if they have a broad outlook. Moreover, the expansion of our horizons is influenced by everything that carries a beautiful content—the sounds of music, poetry, travel, nature with its phenomena and riddles, the process of cognition, the discovery of oneself, and the creation of new ideas and objects with your brain. In addition, all this is interconnected.

It is not in vain, when something pleases us, especially something new, we enthusiastically pronounce the word "beauty"! But beauty requires thought [9].

The life of humanity is developing in such a way that more and more technical objects begin to surround us. Especially in modern big cities, people actually live in an industrial environment. On the one hand, this fact is a direct result of human ingenuity, and on the other hand, in recent decades, we are increasingly beginning to see beauty through a technical lens of perception.

A transformation occurs—the perception of beauty affects ingenuity, and vice versa, ingenuity affects the perception of beauty. Sometimes this transformation proceeds in a dangerous direction, for example, the creation of computer technology more and more involves people in a virtual way of perceiving beauty. The time has come to find balance.

4. Harmony and structural elements in technology

The last century has shown how structural elements in technology affect the development and the life of all humanity. Many technical objects were invented over the entire twentieth century and the beginning of the current century. After the creation of internal combustion engines, people began to gladly produce on an industrial scale and use these devices to solve economic and other problems. However, when a certain level of development was achieved, humanity began to realize the danger that these devices carry in for the environment. So today, humanity is looking for ways to abandon internal combustion engines or, in any case, minimize their use. At the beginning of the twenty-first century with similar enthusiasm, computer technology was received. Seeing in these devices a new stage in the possibility of accelerating industrial and economic growth, we began to use them as extensively as internal combustion engines.

Calculations and organizational capabilities of computer technology, no doubt, allow you to create countless possible models, conduct research and analysis, and develop new programs. However, I like to ask—does the human brain need as much information as can be obtained by computer technology? Alternatively, being energized by computer technology, is our brain capable of harmoniously functioning, fulfilling its natural purpose?

Saturating our brain with a lot of information, there is a danger for people to lose the natural gift to see beauty [10]. Where is the border that technology impairs our ability to perceive and replicate natural beauty? In the author's opinion, computer technology with its extensive capabilities must first study and determine the possibility of a person losing the perception of natural harmony and beauty. Creating a virtual world around him and becoming more and more immersed in it, a person ceases to hear the birdsong and to notice the momentary subtleties of natural phenomena. This narrows his natural visual vision and finally stifles his creativity. A study at the University of Kansas found that young people who backpacked for 3 days showed higher creativity and cognitive abilities [11].

There is a saying that "beauty will save our world." I would add from myself that natural beauty might save the world.

Author details

Levan Torosian
Saint-Petersburg State University of Architecture and Civil Engineering, Saint-Petersburg, Russia

*Address all correspondence to: levantor@mail.ru

References

[1] Kim S, Choe I, Kaufman JC. The development and evaluation of the effect of creative problem-solving program on young children's creativity and character. Journal of Thinking Skills and Creativity. 2019;**33**. DOI: 10.1016/j.tsc.2019.100590

[2] Parsons A. Young Children and Nature. Saarbrucken, Germany: LAP LUMBERT Academic Publishing; 2011. p. 114

[3] Tovée MJ, Swami V, Furnham A, et al. Changing perceptions of attractiveness as observers are exposed to a different culture. Evolution and Human Behavior. 2006;**27**:443-456

[4] Gube M, Lajoie S. Adaptive expertise and creative thinking: A synthetic review and implications for practice. Journal of Thinking Skills and Creativity. 2020;**35**. DOI: 10.1016/j.tsc.2020.100630

[5] Runco MA. Physics. Encyclopedia of Creativity. 3rd ed. 2020. pp. 356-359. DOI: 10.1016/B978-0-12-809324-5.23859-5

[6] Ette O, Lubrich O, editors. Kosmos—Entwurfeinerphysischen Weltbeschreibung. Vol. 1. Frankfurt: Eichborn-Verlag; 1845. p. 17

[7] Lüttge U, Souza GM. The Golden Section and beauty in nature: The perfection of symmetry and the charm of asymmetry. Progress in Biophysics and Molecular Biology. 2019;**146**:98-103

[8] Tiggemann M, Polivy J. Upward and downward: Social comparison processing of thin idealized media images. Psychology of Women Quarterly. 2010;**34**:356-364

[9] Brielmann AA, Pelli DG. Beauty requires thought. Current Biology. 2017;**27**(11):1706

[10] Septianto F, Kemper JA, Choi J(J). The power of beauty? The interactive effects of awe and online reviews on purchase intentions. Journal of Retailing and Consumer Services. 2020;**54**:102066

[11] Louv R, Howard B. Connecting With Nature Boosts Creativity and Health. National Geographic; 2013

Chapter 2

Environmental Art and Wilderness: The Stakes of Beauty

Pascale Weber

Abstract

In the field of environmental art, the concept of beauty is linked to the history of wilderness and Romantism. The beautiful and the sublime arouse a debate that opposes the defenders of a pragmatic ecological commitment against the partisans of a subjective and sensitive creation. In reality, these debates are nothing but old quarrels over the engagement of art, the recognition or rejection of its autonomy towards other disciplines, especially regarding scientific expertise. Approaching creation from the angle of beauty is a way of updating our principles of taste, the ability to judge and also to raise once again what is considered a risk for the individual and for the collective, the exacerbation of the senses, the passions and seduction. Finally, it is about subjecting art to principle of utility that neglects all subjective creation when we are collectively and daily subjected to a regime of guilt and anxiety which orchestrates in turn a form of political and existential resignation. But what remains of aesthetic and artistic experience in such conditions? Does not the rationality of this approach confuse art and communication, denying the emotional connect of art and the fact that beauty is defined by emotional reactions?

Keywords: aesthetic, art, performing arts, beauty, emotion, environmental ethics, perception, presence, representation, romanticism, wilderness

1. Introduction

Like Caspar David Friedrich's famous painting "Wanderer above the sea of fog", the romantic representations of landscapes that have marked the history of Western art bear witness to an immersive experience in rural or wilderness settings, in no human's land. These paintings depict a strange, sometimes hostile setting, escaping from all human power, and in which other forms of animal, plant and magical life flourish. This unknown world arouses strong and contradictory feelings: attraction and repulsion, admiration and fear, serene meditation and morbid fascination. It is necessary to distinguish Romanticism, marked with a capital letter and referring to a nineteenth century artistic movement, to which I have just referred, and romanticism, marked with a lowercase letter, referring to an aesthetic and poetic posture that is not related to a particular era (**Figure 1**).

What is beauty for an artist? There are probably as many answers as there are individuals, I express myself as a performer: for the last fifteen years or so, I have performed within the duo Hantu (weber + delsaux) in non-urban, not very accessible and sparsely populated spaces (Indonesian equatorial forest, Labrador and Canadian Arctic, Northern Norway, Soufriere of Manziana-Italy, Dartmoor-England ...). In these "experiential" works, it is a question of thinking of the place

Figure 1.
Caspar David Friedrich, "Wanderer above the sea of fog", 1817, 98 × 74 cm, oil on canvas, Hamburger Kunsthalle, Hamburg, Germany [Public domain].

from the perspective of the body and conversely of thinking of the body from the perspective of the place in which it is immersed. To describe these performances, I insist on the functioning of our duo: by immersing my body in an unfamiliar space (testing the cold water of a river, a particular brightness or darkness, extreme temperatures, the presence of animals, insects, geographical misapprehension, various dangers and loss of landmarks) I seek to exacerbate my senses and create fictions that reveal a displacement as much as an awareness of my representations of the body and territory. Jean Delsaux records images—photographic or video—that attest to the ecology of our perception revealed by these unusual environments. Thus we work together in the same place and the same context but from two diametrically opposed points of view: the presence of the body in representation and the representation of the body in presence. The feeling of beauty manifests itself for our duo on these two levels: that of presence and that of representation.

2. What is the feeling of beauty?

The feeling of beauty is a construction. It is developed from different emotions (such as fear of danger, appeasement, astonishment, the joy of discovery) that manifest themselves instantly, temporarily and physically (agitation, speeding up the heartbeat and breathing rate, perspiration, etc.) and psychologically (negative or positive thoughts, change of mood).

The feeling of beauty is built from spontaneous emotional reactions. It belongs to each perceiving subject: a performer listening to what the place does to their

Figure 2.
Hantu (#19), "Caldara Di Manziana" (Soufriere/Soufrière), performance n°63, diptych, Italy, 2019. A performing body (a) in a desert place with springs of sulfureous water (b) [©Hantu (weber + delsaux)].

body, or a performer who seeks to represent the experience of presence by putting in motion their gaze but also their own body (reacting to the presence and movements of their partner, which they must accompany with empathy): it is only possible to represent presence by being oneself present to the presence (**Figure 2**).

From the feeling of beauty felt by two people can emerge a deeper emotional construction, implying the overcoming of individual experience and the accession of a new scale of perception. The common experience of a full and indescribable emotional manifestation probably refers to what Kant calls the perception of the sublime. This experience designates a higher level in the graduation of the sensible, I call this feeling of totality aesthetic ecstasy. Aesthetic ecstasy first manifests itself in the encounter between two full artistic experiences, in the subsequent materialization and synchronicity of these two irreducible events, in the framework of a single representation.

Let us clarify. Being present in one's environment consists of an introspective re-centering: it is not a matter of making a "selfie" in an amusement park or on a "remarkable" site along a tourist route. It is often more about locating "neutral" spaces in territories that are already historically or culturally loaded, so as not to have to struggle with the place as a "spectacle". In this space that welcomes emotion, the meditating body then tries to forget the other who is looking at it. Performing presence is to become blind to the image we offer, it is to gradually withdraw oneself from the visible world, to be only present to oneself: to no longer be in representation but to be in one's body, to repatriate the gaze we carry of ourselves from an imaginary external point of view, to become a body of pure sensation. My heart slows down, I hear the blood in my temples beating in rhythm with sounds that I discover, more and more subtle.

This experience is a form of trance and would put me in real danger if Jean Delsaux was not present to monitor the potential risks around me. This shift in attention is necessarily linked to a relationship of trust that confers a sufficient sense of security for my body to deactivate its surveillance functions, to "lose sight" of the surrounding environment. It allows me to focus on the emotional energy manifesting itself in an internal movement of my body. To finish on this point, I add that trusting one's partner first of all means they are the one who sees, lending their eyes, but they are also the one who remembers and constructs the story, lending their memory. Because the particular state in which the performance puts me does not allow me to restore the experience with hindsight. It is the story and image made by my partner I trust to build my own memory afterwards. By changing the values of the scale of sensitivity, part of the immediate reality comes out of the field of our perception: I am no longer more than a set of cells, a constellation of immediate sensations, without intention, therefore without project, that is to say, without memory. When I perform, I lose my identity, I lose all resemblance of what I ordinarily think I am.

Returning to the question, what is beauty for an artist? On the one hand, beauty consists of a harmonious continuity between the body and the space, and on the other hand on a synchronicity between inner experience and the possibility of transmission through the narrative (verbal, iconographic, gestural?) of the feeling inspired by the place in which the body is immersed. The two experiences—presence and representation—are both based on emotions experienced by the body.

2.1 Beauty, the projection of an inner state on a space

"perceiving is an act of attention, not a triggered impression."

James J. Gibson [1]

Perception is an active action, not passive. For the romantic artists of the nineteenth century, the feeling of beauty results from a set of emotions experienced in a body caught in a chaotic environment, in which overpowering natural elements (storms, cataclysms, apocalypse) are expressed, involving a non-human relationship to time (either it unfolds cyclically at the rhythm of the seasons or at the infinite scale of the founding narratives). The contemplation provoked by nineteenth century landscapes is strongly criticized by environmental ecology because the poetics proposed by the experience of the sublime celebrates a wild and authentic nature that could only exist as a fantasy of a lost Eden, and because by ignoring the reality of the industrial world such an aesthetic would be totally inoperative as a form of militant action for the protection of the environment. Landscape, as conceived by Romanticism and as it can still be perceived today, can be seen as a projection of human emotion on a contemplated space: in this sense, it is only a symbolic construction, blind and harmful, arousing and then neutralizing emotions, it has less to do with real spaces threatened by human activity, than with a morbid mental space of consumption and spectacle.

In the romantic posture that prolongs the ideal of the eponymous movement there is indeed a form of meditation that resists pragmatic collective action, just as there is an emotional exacerbation that seeks to subtract the body from a scientific or discursive expertise that is both rationalized and ideological. It is also true that the beauty of the landscape is a projection of an inner state onto a space that stimulates the body and the imagination. Nevertheless, this phenomenological experience is essential for our identity construction because it makes us feel part of a place. Like Realism, Romanticism has developed a sensitive and symbolic lexicon that has changed the way both city dwellers and non-city dwellers see the territory.

However, Romanticism was not so much used to document the reality of nature for its contemporaries, to which specific qualities are conferred, but rather to recognize the lived sensorial experience, and not the theorized experience, of the artist. These representations, the purpose of which is not precisely educational, then infuse all layers of society, inviting them to walk, build and appropriate the landscape.

This point is important, without education about the aesthetic experience of the landscape the state of our roadways would be different: in Southeast Asia for example where I lived as a child and where I went again as an adult, the plastic waste that did not yet exist, is accumulating today, thrown on the side of roads, in the jungle or along the coast as if it were vegetables peels that can decompose. It is representation that changes behavior. Grandiose landscape paintings show the exaltation of the painter confronted with the excessiveness of space. It is a visual shock for the spectator, who thanks to the spectacular scale of the image is invited to reconsider the representation he has of the body, its strength, its ability to act on the course of its existence. In this reversal appears the beauty of the living. And this beauty calls humility. This reversal also shows that the artistic act is only fully realized as a transmissible experience.

Annie Le Brun, a surrealism specialist, speaks about the commodification of art as a "new disfigurement of the world." She considers beauty and ugliness as real political issues, vivid and incisive, that allow us to lead our lives and thoughts with vigilance, when "cynicism makes it possible to neutralize the appearance of true protest" [2]. Oneirism and the supernatural are thus antidotes to anxiety-provoking hysteria, moral guilt and the programmatic modeling of our behavior in the name of the common interest.

The concept of beauty refers to a form of knowledge that is not directly accessible, to which we might—since it is rooted in the intimate and individual experience of the body—blame for being proselyte with difficulty. For this reason working as a duo is important for me, because it rests between the experience of presence and the experience of representation, an unbridgeable gap, even in the case of aesthetic ecstasy, which allows two irreducible experiences to be brought together without being confused as two ascetic regimes, each based on empirical work, separated from the human world, in silence, as two exercises of meditation, vision, revelation. Performing without a public is a permanent re-centering exercise of reviving our expectations, reminding us of what is priceless and preventing our consciousness from being captivated by multiple models: performing becomes a way of dreaming with the body and revealing a world that is not immediately seizable in reality.

Our mental representations are as much affective as they are societal and cultural. They allow us to revisit the infinite complexity of the empirical world. The relationship we develop with the environment and the place we plan to occupy in the biosphere derives from our relationship with knowledge and our ability to see the world around us; for knowledge is first of all a vision. This can be confirmed by classical pedagogical theories involving vision in a form of cathartic knowledge, by the "encyclopaedist" approach of the humanists, or by the passage attributed to the development of information networks from a civilization of language to a civilization of the image. Not only is all knowledge first of all a vision, but also vision is nothing more than knowledge. I would add that any vision cannot be made without a real experience of the landscape, in-situ experience, meditative, immersive experience, before being impressed in our memory by the emotion that this experience arouses.

Beauty is the expression of a total experience, it is a construction and a cognitive representation insofar as it is the image an individual creates of a situation, for example about their presence in a particular place. This representation is at the confluence of sensations and memory: the sensations present are exacerbated by images of past experiences and memory regenerates and feeds on new sensations. According to a cognitivist approach, it is from the image formed in the retina that a first coding takes place, followed by transmission to the brain, from which point a mental representation is generated, upon which symbolic treatments are performed. James J. Gibson criticizes this conception and substitutes it with an ecological approach to perception. According to him one cannot analyze behavior by subtracting the individual from the world, it is necessary on the contrary to consider the being who perceives in relation to their environment and the constraints imposed by this environment.

Beauty is a total experience made up of sensitive knowledge.

In his Manifesto of Romanticism (in the preface to Cromwell), Victor Hugo says that:

> *"Humanly speaking, not everything in creation is beautiful, that the ugly exists beside the beautiful, the deformed near the gracious, the grotesque as the reverse of the sublime, evil with good, shadow with light." [3]*

Hugo goes on to say that "the starting point of religion is always the starting point of poetry. It is all related" [4] which is eventually reiterated by Ernst Cassirer, for whom the human being is essentially a symbolic animal. According to Cassirer, we constantly create symbols to liberate ourselves from pure sensation: to place the body at a distance, to give a word for a sensation in order to exorcise or domesticate it. This is where academic knowledge is unable to find the body that art seeks to achieve. Only artistic language manages to sometimes seize pure sensation. Cassirer dreamed of a general systematics of symbolic forms:

> *"We must, without any prejudice or dogmatic theory of knowledge, seek to understand the particularities of each type of language: scientific language, artistic language, religious language, etc.; we must determine the importance of their contribution to building a 'common world'". [5]*

The meta-symbolic system of Cassirer operates at the meeting point of all spiritual domains: religion, art, science, metaphysics. It proposes a totalizing experience identical to that which constitutes the act of creation. An experience that aims for nothing other than the desire to access and find ways to access spiritualities and modalities of existence that partially evade contemporary thought: the instant of creation is a moment of synthesis of all thoughts and visions—spatial and temporal—that the artist experiences. The artist seeks nothing more than to renew and enrich their sensitive impressions where globalized culture renounces diversity, impoverishing it through the sole concern of maintaining the political order by neutralizing everything that could challenge it and claiming innovations or revolutions are mere reformulations of this order.

2.2 Emotion, perception and sensation

> *"The image is a plant that needs earth and sky, substance and forms"*
>
> Gaston Bachelard [6]

It is through my body that I am present in the world. Immersed in a space that reveals my inner space to me, I wonder what is the nature of the bond that attaches me to my body: what constitutes, nourishes and comes out of both my sensitive knowledge and my perceptible existence. There are different perceptible existences, that of the I and subjectivity, which puts forward a form of pure interiority, and that of being, which aims at a form of universality of existence: a presence of the body to itself. I exist in the presence of my body (**Figure 3**).

It is not a matter of objectifying the body: I can identify with my body to the point of saying I am my body, that is to say that my body is not something external to myself, of which I would have control and ownership and yet I am not only my body. An embodied being is more than a body. I am, I exist only in connection with an action in a given place, even if, in the case of performance, the action sometimes consists of standing still for hours on a branch.

Action refers to a commitment and participation in the world as a body, as part of nature and as a means, in a way that action ends up being confused with existence. One could wonder what drives the artist to continually give birth to new images, to call forth new visions. These representations of nature and the meaning of the body's work in an environment, whether mental or material, directly affect

Figure 3.
Hantu (#12), "The Drift", performance n°23, Plymouth-UK, 2014. Performance supported by the research group for Land-Water and the Visual Arts, the Marine Institute, the Hydrography Group [©Hantu (weber + delsaux)].

our conception of the body, our connection to a territory, and the idea we have of our place among the living. The artist expresses the need to feel a sense of beauty because it is their emotions that weave links of continuity between trivial gestures, performance and various commitments: social, moral, poetic, ideological and spiritual. Through these emotions, the artist confirms their presence and participation in the world that faces them.

Curiously, there are two types of information provided by our visual perceptual system, the first are exteroceptive and they concern the environment, the second are proprioceptive and they concern the moving body. The distinction the Hantu duo has intuitively made between presence and representation seems to have been validated by Gibson, I will develop this point in the next paragraph. Empirically, we have learned to make these two types of information interact by associating the first with the contextual shooting of Jean Delsaux and the second with my physical performance.

Gibson distinguishes the environment of a person (from an animal or a species) that he associates with a point of view (individual or shared) and the private world of an individual related to the biological properties of their perception and to their mode of existence (what is significant for one person is not for another). Perception and point of view would only be adaptive reactions, resulting from conditioning and co-evolution, and from an interaction with their environment. For example, performing in a foreign space is a way of renewing and extending one's own world. Space is not merely an empty container to be filled with matter, but rather overlapping and successive inclusions of places, as time is an intermingling of events that follow and cover each other by introducing modifications in the environment shared by individuals. For the American psychologist, changing the point of view of a person who moves is not an event, but the different stages of the ebb tide, the arrival of the night... modify the environment.

Alphonse De Waelhens tells us:

> *"Most classical theories propose to explain sensation to us by representing our body 'as an apparatus that is at the same time receiver, transmitter, emitter'. However, this image is enough to scramble everything. As soon as the body compares itself to*

> *a wireless telegraph station, sensation is inevitably reduced to a transcribable message. It would be raw data to which I have to make a sign by deciphering it according to a particular code. But this transcription is pure mythology. We never feel that we interpret qualities; we do not see vibrations but red. The notion of vibration comes from scientific explanation and has been developed according to the specific requirements of color physics or the physiology of the senses. It does not correspond to lived experience and plays no role on this level. In addition, the mechanism it implies forces us to radically separate what is felt from the one who feels."* [7]

Yet, precisely the feeling of beauty, without limiting it to the art world, is a construction developed from sensations experienced by a body that has an ability to make itself available to open to what surrounds it. Today, we still have a hard time defining sensations beyond a positivist conception that associates physiological considerations to each of our sensations, by dividing our anatomy into quasi-autonomous sensory organs that are supposed to record and translate information for the brain which manages it. But above all, sensation appears here as an exchange between two objects: sensation (color and its vibration for example) and that which allows for the recording of it: the objectified organ of perception. The experience, the sensitive experience, the felt subjectivity and affectivity of the body are totally forgotten.

2.3 Aestheticization, oneirism and storytelling

> *"In this context, memory is awkward because it assumes a distance with what we live. [...] Yet the computerization of the world abolishes this distance because it constantly seeks the attention of subjects."*
>
> Annie Le Brun [2]

As a form of aestheticization, beauty is reproached for not being in actual contact with the environment, not taking sufficient account of reality, in the urgency of what the experts are telling us of changes in a world in which living species are disappearing and the survival of humanity is threatened, and without any obvious steps being taken to change our way of life. I have opposed this pragmatic conception of "too much reality" that prevents us from changing the order of things, in particular the globalized capitalist logic. The gap with reality is at the very foundation of aestheticization, it enforces a distance, frames our perspective, creates an anticipatory fiction. Without a gap we are condemned to a neutralization of all thought and action: there can be no revolt, no reversal of situation, no change of paradigm. The neutralization of imagination and aesthetic feeling makes censorship useless, all control superfluous, if we understand the ideological stakes, we also understand it leads our society to undergo a consensual force condemning it to inertia.

The gap that art constitutes is the most obvious manifestation of our memorial and imaginative cognitive functions. An artwork is the materialization of a vision of our memory, a memory more or less fantasized (our childhood home...), or a vision of our imagination, arriving in a dream, a day dream, through thought... in resonance or in apparent disconnection with current events. What is important to understand is that we cannot have a perspective on the present or reality, without an aesthetic, cultural, spatial or temporal gap.

Dreamlike beauty is so exploited for the purposes of consumption that we might be tempted to wonder in what spaces of desire and sensation could our dreams unfold today. But why would we be led to believe that in our postmodern societies the dream would have lost its capacity to constitute a gap? For this is how the artistic fictions of our duo come into play. These nocturnal or awakened images intermingle with the images the place of the performance inspire in us. For Philippe Descola, there is no difference between travel stories and fiction. It is a matter of incorporated knowledge, which is a part of the logic of sensible, subjective and concrete qualities at the same time. There is no need to separate the thought experiment of the poet, artist or philosopher from the life experience offered by the curiosity of the traveler, which the different peoples of the world sometimes share with them. Travel stories and fiction tell the story of what keeps people together, today and in different places. For Descola, the dream is the experience of another modality of existence, made of transformation, metamorphosis, and role exchange. The dream is an encounter with the other, and that is what creation intends to offer: creation is the continuation of dream, it takes its source in the dream that it prolongs and allows to last (**Figure 4**).

The waking dream and exchanges with ghosts [8] solicit our faculties of imagination and representation in a unified and multi-dimensional personal cosmogony, unlimited in space or time. From the observation of the divisions and partitions between the different fields of the natural and cultural sciences—which ended up being established at the end of the nineteenth century and which includes anthropology—Philippe Descola proposes an ecology of the other. His research on the different ways of apprehending and perceiving the world, decrypting and describing the environment, allows him to classify and identify, then to recompose an ecology of relations between humans and non-humans:

> *"[...] for each of us, wherever we are, inventing and cultivating modes of reconciliation and types of pressure that can lead to a new universality, open to all components of the world and respectful of some of their particularities, in the hope of warding off the distant deadline to which, with the extinction of our species, the price of passivity would be paid in another way: by abandoning to the cosmos a nature become orphaned by its rapporteurs because they had not been able to grant it any real means of expression." [9]*

"The new universality" that Descola calls for is as reminiscent of Cassirer's ethics, emancipation, and metasymbolic as it is of 'aesthetic ecstasy'. However, in

Figure 4.
Hantu (#5) "Body and Trees/Corps et Arbres", performance n°13, Puy-de-Dôme, France, 2014 [©Hantu (weber + delsaux)].

the sense of beauty where creation is at work there exists a powerfully subversive dimension to which the artist cannot renounce, not even for fear of the extinction of our species: art is the search for a complicity in the refusal of consensus, feeling together from a unique point of view.

For Jacques Darriulat:

> *"To meet, you have to be at least two. At the source of imaginary maieutics, in the atomic moment of astonishment, consciousness intersects the world and the subject intersects the object. [...] Aesthetic encounter is an experience of subversion, and the artwork is engendered by the original coupling of the spirit intoxicated by the ecstasy of sensation and reality which suddenly makes a sign."* [10]

With emancipatory and provocative avant-gardes, the role of art has been to sweep away moral prejudices with violence, to question the very order of things and to confront us with a dark, dizzying nothingness... Have ecological and economic changes along with the amplification of the migration crisis directly and permanently changed the role of art by subordinating it to more pressing questions? Once again in art history it is a question of interrogating the utility of creation and aesthetic feeling: admitting that Ethics is the ultimate form of development for human intelligence that remains for the radicality of the aesthetic experience. As a consequence what sense should I give to the performances we perform in the wilderness?

Influenced by Thoreau, Leopold or Carson, the generation of Naess and Callicot wanted to break with the anthropocentric conception of nature, which manifests itself in a game of projecting the human being through our psychology and aesthetic feelings ... on the space we contemplated. The concept of environmental ethics, formalized in the 1960s and 1970s, emphasizes the moral value of nature degraded by the deregulation of mercantile human activities and defends the principle that from now on any thought or action, artistic endeavor included, should be submitted to this ethical principle. But what place can ethics give to the gap and subversion underpinning aesthetic shock?

I return to the manifesto of Romanticism, in which Hugo poses the stakes of creation:

> *"the modern muse [...] will wonder if the artist's narrow and relative reason must succeed on the infinite, absolute reason of the creator; if it is for man to rectify God; if mutilated nature will be more beautiful; if art has the right to split, so to speak, man, life, creation; if every thing will work better when we have removed its muscle and its spring; if, finally, it is the means to be harmonious more than to be incomplete."* [11]

3. Conclusion

> *"What is priceless are the things that make us live."*
>
> Annie Le Brun [2]

We know the beauty that ethics offers can only be moral when the beauty sought by the artist is anchored in a narrow and relative reason, an inalienable desire submitted to the other. The dialogue that the artist has with the visible is not measured. In its authoritarian developments, dictated by the urgency and the desire for efficiency, the moral conception defended by environmental ethics imposes expertise, specialized policies, an order that both prohibits any holistic approach and substitutes the expression of a consensus for a diversity of languages, which can only weaken our ability to imagine apparatuses that would reinvent political order (and disorder). Creation inscribes a continuity of feelings that ethics will probably judge irreconcilable (beauty/ugliness, attraction/repulsion, adhesion/opposition ...). In a way that the artist suffers from a misunderstanding of their contradictory posture: at the same time marked by great humility and even greater pride.

By allowing opposites to meet, artistic representations and fictions reveal our existential condition. They change the perception we have of our place in the biosphere. There can be no change of thought and behavior without first projecting them into our imagination and body, without assembling them in a cubist collage or in a polysemic narrative of emotions, which would be obstinate, contradictory and paradoxical: synthesis in art is not a simplification with the aim of modeling, it is not coherent, it is the possibility of a heterogeneous and contradictory juxtaposition. Aesthetic expression, feeling, and ecstasy are contradictory experiences, generating both disgust and attraction, fear and anger, admiration and repulsion.

Beauty remains that towards which creation inexorably tends, stubbornly refusing the fragmentation of knowledge. No subversive pretensions of art are justified if the artist must defend the vision of a world cut by the lens of a scientific apparatus of measurement, or defend the objectification and relegation of the human body, quantifiable and transparent, extracted from a sensible world common to all forms of life.

Hence beauty is not, as our mercantile society in crisis claims, a surface reflection, it is the emergence of a deeper reality, affirming the body as subject and not as property, tool or means: I am not my body, the body is the biological seat of all our fabulations, all feelings that allow us to live our human condition, without drying us out or exhausting our desire to live.

Author details

Pascale Weber
University Paris 1-Panthéon Sorbonne, School of Art, ACTE Institute, Paris, France

*Address all correspondence to: pascale.weber@univ-paris1.fr

References

[1] Gibson JJ. The Ecological Approach to Visual Perception. NY: Taylor & Francis, Psychology Press; 2015

[2] Le Brun A. Ce qui n'a pas de prix [What is Priceless]. Paris: Stock; 2018

[3] Hugo V. Manifeste du Romantisme, préface à Cromwell, A. Dupont, Paris, 1828, Manifesto of Romanticism, preface to *Cromwell*, George Burnham Ives (trad.), edited by Little, Brown and Company, 1909: "tout dans la création n'est pas humainement beau, que le laid y existe à côté du beau, le difforme près du gracieux, le grotesque au revers du sublime, le mal avec le bien, l'ombre avec la lumière"

[4] Hugo V. Manifesto of Romanticism, idem, "le point de départ de la religion est toujours le point de départ de la poésie. Tout se tient"

[5] Cassirer E. Logique des sciences de la culture – cinq études, "Perception des choses et perception de l'expression", Paris, Cerf, 1991, The Logic of the Cultural Sciences: Five Studies, Steve G. Lofts (trad.), formerly entitled The Logic of the Humanities. Yale University Press; 1961. p. 123

[6] Bachelard G. L'Eau et les rêves: Essai sur l'imagination de la matière, ed. Corti, 1942, Water and Dreams: An Essay on the Imagination of matter, Edith R. Farrell (trad.) Dallas Institute Publication; 1994. p. 9

[7] De Waelhen A. "La phénoménologie du corps" [The Phenomenology of the Body]. In: Revue Philosophique de Louvain. Louvin, Belgium; 1950. p. 374

[8] Hantu means. Ghost: A spirit, usually of a dead person; apparition; phantom; spectre, specter; spook. Indonesian: Cambridge Dictionary, Dictionary PASSWORD Indonesian-English © 2015 K Dictionaries Ltd.; Available from: https://dictionary.cambridge.org/fr/dictionnaire/indonesien-anglais/hantu

[9] Philippe Descola, Par delà nature et culture, Gallimard, Paris, 2005, Beyond Nature and Culture, Janet Lloyd (trad.). University of Chicago Press; 2013. p. 552

[10] Darriulat J. Principes de philosophie esthétique, "II-Théorie de la connaissance esthétique". Available from: http://www.jdarriulat.net/Essais/PrincipesPhiloEsth/2-TheorieConnEsth.html

[11] Hugo V. Manifesto of Romanticism, op. cit., la muse moderne [...] se demandera si la raison étroite et relative de l'artiste doit avoir gain de cause sur la raison infinie, absolue, du créateur; si c'est à l'homme à rectifier Dieu; si une nature mutilée en sera plus belle; si l'art a le droit de dédoubler, pour ainsi dire, l'homme, la vie, la création; si chaque chose marchera mieux quand on lui aura ôté son muscle et son ressort; si, enfin, c'est le moyen d'être harmonieux que d'être incomplet

Chapter 3

Identity, Self-Identity and Beauty in Chinese Female Worker

Li Yun and Rong Rong

Abstract

In this chapter, we leave "the Cultural Study" in the contemporary researches on beauty, claiming economico-political matters in our self-acceptance and image projection. Rather than studying the standard set for the middle class, we examine the middle-class dream in the self-acceptance of contemporary Chinese female migrant workers. We take "The Sundress", a poem written by Wuxia, a migrant worker in Shenzhen, as an example, arguing that it is mainly the political and economic ideologies that function in the overdetermination of the contemporary working class' projection of beauty in China. In the poem, the "I" speaker, a steam press operator working late in the night, (mis)identifies herself with a middle-class girl who buys and puts on the dress that she irons. She takes great pleasure and tranquillity in picturing the beauty and the leisure time that the idealized "you-I" in the dress enjoys. What's more, our linguistic study testifies that she, as the speaker in the poem, is subjectified by the (mis)identification. Her self-acceptance, activeness and sense of control arise when she overlaps herself with that idealized body image. A great loss, however, is engendered by her having to leave that body and the fantasy world.

Keywords: beauty, identity, the working class, the economico-political, Chinese migrant worker's poetry, overdetermination, subjectification, linguistics

1. Introduction

Contemporary cultural studies on beauty, western or Chinese, besides studies from traditional aesthetic perspectives, can be roughly put into three categories, including research on (i) how mass media represent beauty; (ii) how bodies are regulated, beautified and simulated as the effects of consumerism and mass media; and (iii) whether subversive discourses are plausible or not. In these studies, the representations of women are taken as the most illustrative to display beauty politics. Gender, race and ethnicity are the key notes in case studies, which scrutinize the sexisms and/or racisms in different cultures. What is largely ignored, however, is the class status of the owner of the body. Scholars study mutually middle-class standards but seldom touch the representation of the working class, especially how they perceive and project their own bodily images. It reinforces the public impression that only the upper classes deserve the word "beauty". The academic complies with the global systematic marginalization of the lower classes.

What's more, scholars lose interest in investigating the crucial role of the economico-political in overdetermining (if we could use this Althusserian coinage) the representation of beauty. In China, however, its significance has long been noticed. For example, it dominates the representation of women in Chinese left-wing and socialist cinemas. Because of Marxism, female characters are represented on the basis of class status rather than of sex. Relevant studies notice that body images are moulded in accordance with the ideological and propaganda demands of a particular historical era. In the left-wing cinema (1930s), women are represented as the oppressed, and stars are invited to play charming but wretched "mother" or "daughter" of a beautiful but suffering China. In the early stage of the People's Republic of China (henceforth, PRC) (1949–1965), they are represented as the "master" of "the new nation", and the working-class stand is required. During the Cultural Revolution (1966–1976), female stars perform awakened farmers, workers and soldiers, revolutionaries tearing down "the old world" [1–6]. In the 1980s (the Reform and Opening Up began in 1978), female beauty is also displayed in a way to lead the spectator to reflect on the fate of the nation, especially the economic and political (and of course the cultural) "errors" in the Cultural Revolution [1, 6].

Studies on the presentation of women after the mid-1990s (in 1992 Deng Xiaoping published the southern speech on reform), nevertheless, have "the cultural turn", as consumerism invades Chinese popular cultures. The socialist representation of the Iron Girl during the Cultural Revolution was criticized in the 1990s. What was discarded together with it is the Marxist methodology. Research interests shift to the cultural to detect sexism in Chinese culture. What is missing in these studies, however, is the (re)presentation of working-class women, especially how contemporary female workers accept and display their own bodily images. When we come to the self-perceptions and image projections of the working class, "the Cultural Study" becomes immediately insufficient. Though the cultural is still significant, the weight of the socioeconomic can never be overstated. They function in symbiosis. Often the economico-political trumps the cultural in shaping our acceptance of our identities.

Though their perspectives are middle class, the existing studies do inspire us in one aspect. That is, the acceptance of one's identity and bodily image is always determined by the ideal-I in the imaginary of the owner of the body, as Lacanian psychoanalysis finds. If the physical body does not match the imagined ideal, artificial beautification, physical and/or imaginary, may be needed. Our self-perception and image projection depend on our self-identification. The elements overdetermining our self-identification also determine our image perception. We see, therefore, the limitation of the existing studies on beauty politics. Though elements related to gender, race and ethnicity are not negligible, we must restate the significance of the economico-political in shaping our self-identity and image projection. We don't even need a socialist feminist perspective to inform us that the self-perception of the working class is overdetermined by their economic conditions as well as how they are interpellated or hailed by the ideological apparatus.

In this chapter, therefore, we study the self-perception and image projection in the poetry of contemporary Chinese female migrant workers. Here we see how their socioeconomic status and the way they are hailed shape their acceptance of their bodily images. We limit our study to a linguistic analysis of the projection of beauty in "The Sundress" [7], a poem written by Wu Xia, who worked in a garment factory in Shenzhen when she wrote it in 2013. The poem describes vividly the interior monologue of a female worker who works late in the night, ironing a beautiful sundress. The "I" speaker is talking to "you", another female figure in her imaginary, fantasizing how "you", putting on the dress "I" ironed, can enjoy leisure time in some park with your lover. "I", the worker, (mis)identify myself with that idealized middle class "you-I" [8] and pictures a pretty body that "I" should have.

We argue, however, that the misidentification is paradoxical in the Chinese context. It displays a reverse of the acceptance of the working class of their social identities, caused by the transformation of economic conditions and political ideologies. In socialist China, they are addressed as the master of the nation and the builder of its economy. In the post-socialist era, however, they lose their social privileges, politically and financially. Their self-acceptance changes as their social status changes. To understand it, we can compare the self-projection in Wu's poem with the (self-)presentation of working-class women from 1949 to 1964. As Wang Zheng shows in her researches, in the socialist epoch when working women are interpellated as the glorious working class, female film directors and magazine editors present healthy, strong and working women as beauty. The sick image of "bourgeois ladies" is to be transformed [5]. In the twenty-first century, however, the class status of the working class "falls". Migrant workers are regarded as "strangers" or "outsiders" (外来) by the cities in which they live and work. Economic ideologies belittle their contributions to the nation. At the same time, they become financially vulnerable [8–11]. In short, the working class loses its pride in self-acceptance, while middle-class standards become global.

We see therefore the middle class dream the imagined ideal-I in "The Sundress". In the poem, there is a sharp contrast between two bodily images. One wears a sweaty factory uniform, working late in an enclosed factory and having no time and money to enjoy the prettiness and happiness brought by the dress she makes. The other puts on the dress, having a wonderful time of love in a park. The imagined body is active, running, swirling and laughing, glowing in the dress that "I" made. The real body that "I" am in, however, is confined to my position in the packing area of the factory, busy ironing and packing the dress that "you" wear. The imagined charm of "your" body is in sharp contrast with the plainness of "my" body in reality. I have great pleasure in picturing the beauty in that body because I take subconsciously that body as an ideal-I that should be me. Therefore, at the end of the poem, we sense a loss in her when she detaches herself from the fantasy since she has to pack the dress so that it could go to some boutique to be bought by the "girl unknown" before she gets off work.

We see the effects of consumerism and mass media in this idealization, because the "beauty" represented by "you" exemplifies the middle-class tastes that "I" may get from mass media. What's more significant, however, is the (mis)identification in the fantasy. "I" take "your" life as the ideal life, "your" image as the ideal image, "your" body as the ideal body that "I" should have. I project all my fantasies about beauty onto that image, regarding it as an idealized "my" body and describing vividly how that body looks. On the other hand, the "I" does not take her working body as pretty. Rather, she believes that the body of the ideal "you-I" is glamorous. The glamour comes partly from the dress I iron but more from the fact that you enjoy the dress and the leisure that I cannot afford. I take that middle-class you as superior to me. I want to please that "you-I". In contrast to the foregrounded charm of "your" body, however, "my" body seems invisible. "I" seldom focus on it. The reader only sees a pair of hands and a vague figure in a factory uniform. The vagueness betrays an unsatisfaction in the speaker with her identity as a migrant worker.

Our linguistic study on the poem testifies the (mis)identification. We argue that "I" become active only when "I" am connected to "you". Conventionally speaking, the "I" is the subject in the poem because she is the first-person narrator and is in an interlocution with someone in her imagination. We, however, do not read much activeness in this subjective position. On the contrary, our linguistic study discovers that her activeness comes from her identification with "you". It is this (mis)identification that makes "me" the subject. In the poem, the "I" seems passive before she enters her fantasy world and after she detaches herself from it. But when she is in her fantasy and mixes herself with the imagined body, she becomes active, both physically and mentally. Not only the idealized body is active, but the real body in parallel also becomes active and capable of control. Her imagination is activated. She gives a more vivid and specific description of the movement of her real body, the working hands. The "I" speaker is subjectified by her misidentification with the idealized "you-I". We can illustrate it through a close reading of the poem.

2. Plainness in self-identification

But before the reading, we need to know some background information about the poet. Wu Xia (邬霞) was born in Sichuan province, China, in 1984 but has been working in factories in Shenzhen since the age of 14. As a factory worker, she has been writing online for two decades. She is known for her faithful representation of factory life and for the mixing of reality and dream in her poems. "The girls under Wu Xia's pen live a better life than hers. In 'The Sundress', we don't read the tragic and the repression in most poems written by migrant workers".[1] Both tendencies are brought to the fore in "The Sundress", which is collected in *The Verse of Us* (2015). The poem also exemplifies Wu's preoccupation with a range of mental and emotional experiences of being a female factory worker in contemporary China, particularly her desire of escaping factory life and of living a better life. We see the desires in her creating and self-identifying with another "me" in the poem.

1 包装车间灯火通明 The packing area is flooded with light
2 我手握电熨斗 I hold the iron
3 集聚我所有的手温 Collecting all my hand's warmth
4 我要先把吊带熨平 I am going to press the straps flat
5 挂在你肩上不会勒疼你 So that they won't dig into your shoulders
6 然后从腰身开始烫起 And then press up from the waist
7 多么可爱的腰身 A lovely waist
8 可以安放一只白净的手 Upon which a fine hand can lay
9 林荫道上 And on the tree-shaded lane
10 轻抚一种安静的爱情 Caress a quiet kind of love
11 最后把裙裾展开 Finally I'll smooth the dress out
12 我要把每个皱褶的宽度熨的都相等 To iron the pleats to equal widths
13 让你在湖边 或者草坪上 So you can sit by the lake or on a grassy lawn
14 等待风吹 And wait for a breeze
15 你也可以奔跑 You can also run, but
16 一定要让裙裾飘起来 带着弧度 You must let the dress fly with curve
17 像花儿一样 Like a flower
18 而我要下班了 But I'm getting off work
19 我要洗一洗汗湿的厂服 I need to wash my sweaty factory uniform
20 我已把它折叠好 打了包 I have folded it packed it
21 吊带裙 它将被装箱运出车间 The sundress It will be taken out of the factory
22 走向某个市场 某个时尚店面 To some market Some boutique
23 在某个下午或晚上 Waiting in some afternoon or evening
24 等待唯一的你 Only for You.

[1] "Contemporary Chinese Workers' Poetry: Speaking the Inside Story" (http://www.xinhuanet.com/politics/2015-02/06/c_1114279855.htm, accessed: February 1, 2020). Migrant workers' poetry writing has been a cultural phenomenon in contemporary China. It is estimated that 20,000–30,000 peasant workers write poems in their blogs, QQ or on literary websites. In 2015, *The Verse of Us* was published, causing public attention. Qin Xiaoyu, the editor, also made a documentary of the same title, demonstrating the life and poetry of contemporary Chinese workers. In it the invited worker poets read their poems. Wu Xia reads "The Sundress" (http://www.docuchina.cn/2016/11/21/VIDEoulgdZbeLxxBCDTrS1go161121.shtml, accessed: February 1, 2020).

25 陌生的姑娘 The girl unknown
26 我爱你 I Love you

The poem consists of 26 lines which feature the voice of a clothing factory worker, who talks about her work as a steam press operator. Her job is to iron pieces of sundress before packaging. The speaker also describes her emotional involvement in the process of ironing the dress, represented in the form of her addressing a hypothetical hearer "you" about how the dress should be worn (we will explain below why we treat the speaker as female). The poem is divided into four stanzas; the first three describe, respectively, the scenes before ironing, during ironing and after ironing, which correspond to the time before her fantasy, in her fantasy and after her fantasy.

Of course, she feels pretty in her fantasy. Our study, however, is not on what that prettiness is but what causes that charming feeling, i.e. her self-acceptance and sense of control. Remarkably, the linguistic features in the poem show that her activeness and sense of control fluctuate as she enters and leaves the dream world. Her self-acceptance also rises when she prepares for the ironing and fantasizing. It goes to the peak when she invests herself into the fantasy and overlaps herself with the ideal image. When she has to leave the dream world, however, she loses her activeness, self-acceptance and sense of beauty. In the last stanza, her language indicates a great reluctance to be detached from the idealized image.

The "I" speaker only feels active and charm when she is not in her real body. The poem begins with plain language, describing the reality that she is in. But when she gets closer to the dream world, her language becomes more vivid and colourful. The first few lines establish the physical environment where the speaker is working:

1 包装车间灯火通明 The packing area is flooded with light
2 我手握电熨斗 I hold the iron
3 集聚我所有的手温 And collect all my hand's warmth to it

One of the first and most prominent features in these lines is the specificity of the here-and-now context in which the speaker is situated. This context not only involves the specific place, time and participants but also is "coloured" by a certain mood of commitment. First, the poem begins with the locative expression "the packaging area" which has the duo effect of anchoring the ongoing events in a particular place and also activating readers' frame knowledge about factory work. With this knowledge, readers are able to interpret the word "iron" in line 2 as specifying "I" as a steam press operator and also read the modifier "灯火通明" (literally meaning "fully lit up") as suggestive of the time of the day (i.e. late at night).

Also, the adjective "所有的" ("all") modifying "my hand's warmth" is used to indicate the high degree of temperature caused by the pressing work. Both these adjectives are used as amplifiers ([12], p. 429) which function to dramatize a tough working environment against which the speaker is about to begin her night shift. Against this setting, the speaker is presented, however, as physically static. The first verb "握" ("hold") in line 2 is a transitive stance verb, suggesting no change for the time being ([12], p. 747), while the second verb "集聚" ("collect") in line 3 is used metaphorically to refer to the accumulation of heat from my hands and the transmission of it to the iron (so that I could iron the dress with love). Altogether, these introductory lines help to position the reader within an immediate and close distance with the speaker who, while being physically static, is fully committed and prepared for the pressing work.

3. Beautification in misidentification

The impression that the speaker is fully prepared and committed in her work is reinforced by lines 4, 6, 11 and 12 which describe a series of actions that constitute the whole work of steam-pressing:

4 我要先把吊带熨平 I am going to press the straps flat
6 然后从腰身开始烫起 And then press up from the waist
...
11 最后把裙裾展开 Finally I'll smooth the dress out
12 我要把每个皱褶的宽度熨的都相等 To iron the pleats to equal widths

The language describing the speaker's planning contains a number of linguistic features that are typically associated with making commitments. To begin with, there are four "把" (*bǎ*) in lines 4, 11, 12 and 20. The word often appears in what Chinese grammarians call the disposal construction, i.e. *bǎ* +NP+VP. The verb itself has been highly grammatized and acts only as a disposal marker instead of a verb, and the following NP+VP structure is often considered as a resultative complement that gives prominence to the causative result the object (recipient) receives from the subject (actor) ([13], p. 150, [14], p. 99). In Wu's poem, the *bǎ* +NP+VP structure serves the function of emphasizing the possible results that the speaker intends to achieve.

In addition, the content of the plan is characterized by a detailed list of the particular parts of the dress ("straps", "waist", "pleats") and a series of adverbials denoting sequence ("first", "then", "finally"). Cognitive scientists and linguists studying narrative comprehension treat these linguistic devices as world builders that help readers mentally construct the story world depicted in a stretch of narrative text [15–18]. The more detailed a story world is described, the more concrete and vivid this world would be conceptualized by the reader. As such, expressions including temporal and spatial deixis and nominal and verbal references suggesting people, objects and actions play an important role in highlighting the scenic vividness [16] as well as encouraging empathy from the reader [19, 20].

There is, however, a contrastive effect of under-specification evolving from the process of plan-making. In particular, this comes from the description of a series of imagined situations resulting from the speaker's planned actions:

5 挂在你肩上不会勒疼你 So that they won't dig into your shoulders
...
7 多么可爱的腰身 A lovely waist
8 可以安放一只白净的手 Upon which a fine hand can lay
9 林荫道上 And on the tree-shaded lane
10 轻抚一种安静的爱情 Caress a quiet kind of love
...
13 让你在湖边 或者草坪上 So you can sit by the lake or on a grassy lawn
14 等待风吹 And wait for a breeze
15 你也可以奔跑 You can also run, but
16 一定要让裙裾飘起来 带着弧度 You must let the dress fly with curve
17 像花儿一样 Like a flower
...

One of the prominent linguistic features in depicting the speaker's imaginations is the introduction of "you" which invokes an illusionary addressivity. Despite the fact that the second-person pronoun has the effect of bridging the spatiotemporal divide between the two interlocutors, the hearer referred to as “you” is hypothesized by the speaker and exists only in her future-related imagination. This understanding allows readers of the poem to interpret “you” as an anonymous, generic figure whom the narrator has constructed herself based on her knowledge and belief.

In narrative comprehensive research, readers tend to first evaluate the trustworthiness of the narrator ([18], p. 81), which is done based on the readers’ knowledge of the narrator, as well as on their perceptions of the narrator’s level of knowledge. This is termed as “belief world” which consists of a package of knowledge and beliefs that are *only* related, and hence plausible, to this particular character [17, 18, 21]. In “The Sundress”, the speaker is viewed as a reliable narrator whose personal knowledge concerning steam-pressing is recognized by her occupation. However, as the speaker continues to construct her future projection based on her own false belief or strong desire, the speaker’s credibility begins to decline—interestingly, this process is indicated by a list of linguistic features which create the tone of uncertainty.

The first characteristic feature suggestive of under-specificity includes a series of unprecise “world builders”. There are altogether three hypothetical scenes, line 5 describing the illusionary addressee “you” putting on the dress, lines 7–10 going on a date and 13–17 being outdoors. The spatiotemporal frames for these imagined scenes are unprecise. For example, the locations in the latter two hypothetical scenes are specific in terms of type (tree-shaded lane, by the lake, on the grassy lawn) without giving specific information concerning exactly where. Moreover, the references for the imagined events are indefinite, suggested by the use of generic quantifiers “一只” in line 8 and “一种” in 10. All these features add up to create an effect of under-specification: the speaker seems to be engaged actively in constructing and experiencing an imagined yet unspecified fantasy.

Meanwhile, this effect is foregrounded by the contrastive effect of specificity which is generated by the same lines of the poem depicting the actions “you” have been engaged in. There are three verbs denoting “you” as an active participant (actor) in a material intention process (“wait” in 14, “run” in 15, “let” in 16) and three verbs denoting “you” as the passive recipient in the material event processes (“hang” in 5, “place” in 8, “caress” in 12). Material process is a term from systematic functional linguistics (SFL) to describe the roles of the subject and the object in a sentence and their relationship. Material intention process refers to the process in which the actor has the deliberation in performing the action and is commonly associated with a goal, whereas material event process does not necessarily have a subject (or actor) since its focus is on the goal or on the object that is being affected by the process. In the poem, the material intention processes mentioned above highlight “you” as an active actor engaging in a series of outdoor activities, all of which are indicative of freedom and happiness. The material event processes (“hang”, “place”, “caress”) treat the participant “you” as the receiver of all gentle actions, thereby foregrounding the sense of tenderness and tranquillity.

In addition to the vividness, the imagined scenes are also “coloured” by a tone of positivity. It is interesting to note, firstly, that the imagined lover, referred to by means of metonymy as “a hand”, is modified by the adjective “白净的”. The literal translation is white (“白”) and clean (“净”). A quick search of the 780 million-word CCL corpus (developed by the Center for Chinese Linguistics, Peking University) shows that the expression is associated with both female and male. The male-associated connotation suggests either neatness in appearance or physical weakness. In the poem, the collocation focuses on the good physical state of a male’s hand as contrastive to a hand that is dark and dirty due to heavy labour work. In terms of discourse analysis, this value-laden adjective is viewed as an indicator of the speaker's "world-view" ([22], pp. 130–134) or "ideological viewpoint" ([23], p. 277) about social superiority of "you". The ideological assumptions of this description are carried by further an abundance of positively charged adjectives including "lovely", "fine", "tree-shaded" and "quiet", which help create a first-date scenario that is correlated with the ideological associations the speaker has for the middle class ([8], p. 60).

The set of ideological assumptions discussed above tend to foreground (i) the speaker's belief world that it would be the middle-class girls who can buy and wear the sundress and also (ii) a wish world that she could be one of them. It is arguable that the specificity of the imagined scenes is associated with the speaker's wish in a "you" that can realize all her fantasies by wearing the sundress, whereas the effect of vagueness is contributed by the fact that the speaker's limited amount of knowledge about how these scenes might be played out in real life (i.e. *prospective extension of a character's knowledge-world*, 21: 116). In general, the high degree of idealization of these hypothetical futures in turn implicates a mixed attitude of unsatisfaction, compromise and hope that the speaker holds towards her real-life situation.

This attitudinal combination of belief and wish is supported by the combination of boulomaic, deontic and epistemic modalities. All the features identified so far contribute to vivid representations of the steam-press work and the dress-wearing that do not take place in reality by the speaker. For the dress-wearing scenes, the unrealized status of the actions is implied by the modal verbs "可以" ("*kě-yǐ*") in 8 and 15 that express not only the speaker's notion of possibility (paraphrased as "it is possible to") but also her permission (paraphrased as "be allowed to"), thereby also conveying some degree of intrinsic human control over the hypothetical situations ([12], p. 219, [18], p. 99).

The sense of control is further intensified by the modal verb "一定要" ("yī-dìng-yào"; literal meaning "must") in line 16. Similar to the English modality, the Chinese expression "yī-dìng-yào" expresses the intrinsic modality of duty while conveying the extrinsic sense of necessity. In the poem, from "you can run" to "you must let the dress fly", the speaker apparently maximizes her control over the imagined addressee, a process which could be interpreted as the speaker's increasing involvement in constructing a hypothetical world, alone with which also her increasing desire in living in this fantasy world.

4. Loss in demisidentification

Whereas the dress-wearing scenes are supported by the speaker's strong desire, the dress-pressing scene can then be argued to be in line with her sense of determination. Different from the modal verbs used for depicting the dress-wearing situations, the work scene is mainly delivered through the modal verb "要" (yào), first in lines 4 and 12 and then in 18 and 19. According to the *Modern Chinese Dictionary* [24], three senses are mostly relevant to the understanding of "yào": (i) "to want, to wish"[2]; (ii) "being determined to"; and (iii) "being going to". The verb "yào" in lines 4 and 12 is conceptualized based on a combination of the last two senses, denoting a semantic mixture of planning and determined willingness to carry out the plan. This is similar to the modal verb "will" in modern English which has an

[2] According to grammarians (Po-Ching & Rimmington, 2003, p. 284), "yào", when denoting wish, tends to suggest a strong desire in comparison to the modal verb "xiǎng" (lit. want) indicative of a lighter degree of desire. Also, it is the first sense of "desiring" that is used in the published version.

overlapped meaning of volition (intrinsic) and prediction (extrinsic)" ([12], p. 219).

The same modal verb "yào" underlined in lines 18 and 19, however, suggests a different sense from the previous one in lines 4 and 12, thereby signalling a change of attitude:

18 而我要下班了 But I'm getting off work
19 我要洗一洗汗湿的厂服 I need to wash my sweaty factory uniform

The first thing to notice in the above lines is the connective "ér" (literal meaning "but; however") that is associated with the notion of contrast. Prior to these lines, the speaker was described as being actively and closely engaged in her fantasy. Her engagement with the fantasy, parallel with the steam-pressing work in reality, is now drawn to an end as she ends her work shift. Also, the sense of authoritarian control resulting from her fantasizing, however, has been challenged since the sundress is to be sent away and that she has to draw herself out of the fantasy. Therefore, the connective "but" is indicative of a tone of loss, which in turn helps us to interpret the modal verb "yào" in 18 and 19 as an expression of self-obligation (paraphrased as "have to get off work") in which the speaker appeals to her own sense of duty, and this "duty", apparently, is driven by the external force, i.e. the fact that she does not own the dress.

From this point onwards, the "tone" of the poem starts to change, from an intimate and engaging kind to the one with distance, loss and lack of control. This is manifested linguistically through the following features:

20 我已把它折叠好 打了包 I have folded it packed it
21 吊带裙 它将被装箱运出车间 The sundress It will be taken out of the factory
22 走向某个市场 某个时尚店面 To some market Some boutique
23 在某个下午或晚上 Waiting in some afternoon or evening
24 等待唯一的你 Only for You.

First, the reference to the object "sundress" changes from the second-person pronoun to the inanimate third-person pronoun "it". In terms of deictic shift theory, the change of pronoun signals the speaker's perceptual shift in relation to the sundress, as becoming more distal in terms of attitude. This perceptual shift is also consistent with our previous interpretation of lines 18–19 that the emotional bounding between the speaker and the sundress was broken by the fact that the speaker has completed her work in relation to the dress. Even though the pressing worker has devoted her time, effort and even emotion to the sundress, the latter, yet still, leaves her behind. The sense of loss is further reinforced by the multiple uses of indefinite references such as "某个" (literal meaning as "some") and the connective "or". From these references, it is clear that the speaker has no idea what would happen to the sundress she has pressed nor has the idea as to who would buy the dress.

Nevertheless, the reader can sense her unwillingness to detach herself from the fantasized "you-I". It is interesting to note that the buyer of the sundress is characterized by concrete referencing of "you" in line 24. So even the text world switches to a future world that is beyond the speaker's knowledge, i.e. a hypothetical future world that the speaker still tries to place the imagined "you" inside of it. Therefore, lines 21–24 are arguably still a presentation of hypothetical future world based more on the speaker's fantasy than on her knowledge.

25 陌生的姑娘 The girl unknown
26 我爱你 I Love you

In the last two lines, two textual features interrelated with one another deserve our attention. The first to notice is line 26 which writes "I love you". This sentence, in relation to the whole poem, serves as a register signaling the literary convention of a love letter. The researchers of this paper have conducted a seminar, surveying the students' readings of this poem. After their first reading, three out of eight postgraduate students majored in literature and linguistics have identified, based mainly on line 26, the speaker as a male and the poem a love letter. Interestingly enough, it is also this kind of taken-for-granted identification that evokes a schematic deviation within these two last lines of the poem.

Her wording, however, indicates that despite her reluctance to accept it, she is aware somehow that the fantasy world is leaving her. Conventionally speaking, we should have a good, if not full, knowledge about the person whom we claim to love. In the above lines, however, the love is associated with the person whom the poetic speaker refers to as "the girl unknown". It is odd to refer to someone we have attached such a strong emotion to as being simultaneously specific ("the only" in 25 and "the" in 26) and unspecific ("unknown" in 25). The only interpretation from this contrastive pair of references is that the sense of specificity is related to the hypothetical enactor constructed by the speaker's fantasy, while the non-specificity results from the limited knowledge the speaker has in relation to her fantasy.

Similar to all the paradoxical connotations described above, the "girl", projected by the strong desire and a close identification from the speaker, is however, underspecified due to the speaker's lack of real-life experience. There does exist an actual future domain in which *some* girl would buy the sundress, and when this happens, the actual "you" would not overlap with the hypothetical "you". We hence argue that the paradoxical contrast between "I wish to become" and "I know how to become" further reinforces the poetic theme of loss.

In short, she is reluctant to leave the fantasy world because it is a great loss. She is losing the most important part of her. She knows it but she also knows that any effort to cling to it is in vain. We see in this (mis)identification her desire to change her class status. It is a reverse of her identity as the working class. And as we have pointed out above and elsewhere, this reversal is caused by the "fall" of the working class in the economic and political system in contemporary China. The economico-political plays an important role in shaping her sense of beauty and self-acceptance.

Author details

Li Yun* and Rong Rong
School of Foreign Languages, South China University of Technology, China

*Address all correspondence to: flly@scut.edu.cn and flrrong@scut.edu.cn

References

[1] Cui S. Women Through the Lens: Gender and Nation in a Century of Chinese Cinema. Honolulu: University of Hawai'i Press; 2003

[2] Hu J. Projecting a Nation: Chinese National Cinema Before 1949. Hong Kong: Hong Kong University Press; 2003

[3] Zhang Y. Chinese National Cinema. New York and London: Routledge; 2004

[4] Roberts R. Maoist Model Theatre: The Semiotics of Gender and Sexuality in the Chinese Cultural Revolution (1966–1976). Koninklijke Brill NV: Leiden and Bostom; 2010

[5] Wang Z. Finding Women in the State: A Socialist Feminist Revolution in the People's Republic of China, 1949-1964. Berkeley, CA: University of California Press; 2016

[6] Lu S, editor. Transnational Chinese Cinemas: Identity, Nationhood, Gender. Honolulu: University of Hawai'i Press; 1997

[7] Wu X. Diaodaiqun (The Sundress). In: Qin X, editor. Wo de shipian (Verse of Us). Beijing: Zuojia chubanshe; 2015. pp. 327-328

[8] Li Y, Rong R. A middleclass misidentification: Self-identification in the autobiographical poetry of Chinese female peasant workers. Positions: Asia Critique;**27**(4):55-76

[9] Yu J. Anyuan shilu: yi ge jieji de guangrong yu mengxiang (Anyuan: The Glorious Dream of a Class). Jiangsu Renmin Chubanshe: Nanjing; 2011

[10] Pun N. Made in China. Durham. NC: Duke University Press; 2005

[11] Yan H. New Masters, New Servants. Durham and London: Duke University Press; 2008

[12] Quirk et al. A Comprehensive Grammar of the English Language. London: Pearson Longman; 1985

[13] Po-Ching Y, Rimmington D. Chinese: An Essential Grammar. London: Routledge; 2003

[14] Shi Y. Hanyu yufa (Chinese Grammar). Shangwu Yinshuguan: Beijing; 2010

[15] Van Dijk TA, Kintsch W. Strategies of Discourse Comprehension. New York: Academic Press; 1983

[16] Emmott C. Narrative Comprehension: A Discourse Perspective. Oxford: Oxford University Press; 1997

[17] Werth P. Text Worlds: Representing Conceptual Space in Discourse. Harlow: Longman; 1999

[18] Gavins J. Text World Theory: An Introduction. Edinburgh: Edinburgh University Press; 2007

[19] Whiteley S. Textual world theory: Real readers and emotional responses to The Remains of the Day. Language and Literature;**20**(1):23-42

[20] Johnson-Laird PN. Mental Models: Towards a Cognitive Science of Language, Inference, and Consciousness. Cambridge, MA: Harvard University Press; 1983

[21] Ryan ML. Possible Worlds: Artificial Intelligence and Narrative Theory. Bloomington and Indianapolis: Indiana University Press; 1991

[22] Fowler R. Linguistic Criticism. Oxford: Oxford University Press; 1986

[23] Short M. Exploring the Language of Poems: Plays and Prose. Harlow: Longman; 1996

[24] Lv S et al. Xiandai Hanyu Cidian (Modern Chinese Dictionary). 2nd ed. Shangwu Yinshuguan: Beijing; 1994

Section 2

Formal Science, Natural Science and Beauty

Chapter 4

Beauty in Mathematics: Symmetry and Fractality

Vladimir A. Testov

Abstract

The most important concepts underlying beauty are the concepts of symmetry and fractality, but the relationship of these concepts has not yet remained clear. For centuries, beauty was understood only as a stable order and symmetry. Synergetic worldview allows us to give a new assessment: beauty can be seen as an attractor, the result of self-organization of nature, or the flight of human thought. On the one hand, fractality can be considered one of the manifestations of symmetry in an expansive sense. On the other hand, symmetry can be considered a manifestation of fractality with a finite number of iterations. Thus, the concepts of symmetry and fractality are closely interrelated. Symmetry reveals in beauty a stable order, and fractality reflects in beauty the result of the self-organization of the chaos of nature or the freedom of human thought. Symmetry and fractality are two opposites, mutually complementing each other, aesthetically and mathematically mutually passing into each other. Thus, symmetry and fractals are the most important concepts for the disclosure of the beauty of the universe, which determine their importance for learning. The concept of self-similarity can serve as a basis for acquaintance with fractals.

Keywords: symmetry, self-similarity, fractality, chaos, order, golden ratio, Fibonacci numbers, neutrosophy

1. Introduction

Since ancient times, it has been realized that the beauty of the world exists independently of human consciousness. The sense of beauty is a product of reflection in human consciousness of really existing aesthetic properties of objects. Thanks to this feeling, a person develops an aesthetic culture, the ability to create, and an idea of the beauty of the surrounding world is formed. The beauty of the universe is revealed in a special way in every science. Mathematics is not only a coherent system of knowledge and tasks, but also a unique means of understanding the beauty of the world. Studying mathematics, a person discovers new fragments of beauty, moving to the understanding and then to the creation of beauty and harmony.

Beauty confirms and complements the universality of mathematical theorems and formulas that work equally effectively in living and inanimate nature, in atoms and in the universe, in scientific discoveries, and in works of art. Beauty helps to accept the world around with admiration, and mathematics makes it possible to realize the perceived phenomena and objects and deepen knowledge about the harmony of the world.

IntechOpen

Scientists have proved that behind various artistic, architectural, linguistic, and musical creations and natural phenomena, there are general mathematical regularities. Special attention is paid to symmetry, the Golden ratio, and in recent decades also to fractality. The process of human cognition of beauty, self-organization of his knowledge, is a very complex, yet little studied procedure. In this process, the scientific concepts of symmetry and fractality are decisive.

Symmetry is an ancient universal symbol, which from generation to generation forms in the consciousness of man the idea of harmony of the universe. The awareness of the order and beauty of the universe is the meaning of symmetry in science, and the meaning of symmetry in art is the product of beauty and perfection. In culture, the idea of symmetry goes from a visual natural symmetry to a scientific concept. Currently, the concept of symmetry is widely used in a variety of sciences: physics, chemistry, crystallography, psychology, etc.

2. Literature review

A large number of works are devoted to beauty in mathematics. The most fundamental works in this direction are the books of Voloshinov [1], Varga et al. [2], Tarasov [3], etc. A number of scientists-mathematicians tried to reveal the concept of beauty of a mathematical object, paying special attention to the presence of a measure of order in a mathematical object. In particular, Henri Poincaré noted that "he sees the properties of beauty and grace in the elements, harmoniously arranged in such a way that the mind can easily capture them completely, guessing the details." In his opinion, the system of mathematical knowledge brings order symmetry, understood as the harmony of the individual components of this system, their happy balance, giving its components an internal meaningful unity [4].

According to V.G. Boltyansky, "the beauty of a mathematical object lies in the presence of isomorphism between the object and its visual model, the simplicity of the model and the surprise of its appearance, which can be briefly written in the form of the formula: beauty = isomorphism + simplicity + surprise" [5].

The most clear formula for the beauty of a mathematical object was defined by Garrett Birkhoff: $M = O/C$, where M is a measure of the beauty of the object, O is a measure of the order in the object, and C is a measure of the effort expended to understand the essence of the object [6].

In accordance with this, the most common view, the beauty of the object will increase as the ordering of its structure. From this point of view, the most obvious form of order in nature and human creations is symmetry.

According to G.I. Sarantsev, the following can be attributed to the signs of beauty of a mathematical object: "the correspondence of a mathematical object to its standard, stereotypical image; order, logical rigor; simplicity; universality of the use of this object in various branches of mathematics; originality, surprise" ([7], p. 15).

The ancient Greek philosophers called the order in the universe as cosmos, which was opposed to chaos. They combined in the concept of "cosmos" two functions – the ordering and aesthetic functions.

The most important concept, since ancient times underlying beauty and harmony, is the concept of symmetry. According to the prominent German mathematician of the twentieth century Hermann Weyl "Symmetry… is the idea by which man for centuries tried to comprehend and create order, beauty and perfection" ([8], p. 37).

The concepts of symmetry have existed among many peoples since ancient times, but in a broader sense as the ideas of balance and harmony. The principle of symmetry was recognized as one of the foundations of the classical scientific picture of the world, especially after, in 1918, Emmy Noether proved the famous theorem that every continuous symmetry of a physical system corresponds to some law of conservation.

The main characteristics of the concept of symmetry are proportionality, commensurability, and invariance manifested in any transformations (**Figure 1**).

But in nature, parts like each other cannot exactly coincide, so the symmetry in nature is never absolute (**Figure 2**).

The concept of symmetry in mathematics, on the contrary, reaches absolute rigor of definitions. In particular, in geometry, symmetry is the ability of shapes and bodies to retain shape and properties under some transformations. Even at school, students are introduced to mirror and central symmetry. In addition, there is a rotary symmetry, which means that when the body rotates in space at some angles,

Figure 1.
Taj Mahal.

Figure 2.
Elbrus.

its appearance will not change. Considered and other types of symmetry are unlike the usual: translational (repeated by some rule pattern through the same or regular distance, for example, the pattern on the wallpaper, parquet, lace, and tiled roof, and pattern on the skin of the snake), color (mirror reflection with a change of color, for example, chess pieces arranged in the same order), screw (observed in the arrangement of leaves on the stems of many plants, so that they do not obscure each other from the light; another manifestation—the device scales pineapple).

In his book, Hermann Weyl understood symmetry as the immutability of any properties of an object under some kind of transformations. These transformations may not only be movements. H. Weil devoted one of the chapters of his book to ornamental symmetry. In patterns and ornaments, orderliness and subordination to a certain set of rules can also be found. In the case of potentially infinite patterns, as H. Weyl notes, "the operation with respect to which this pattern remains unchanged does not necessarily have to be a movement, it can also be a similarity" ([8], p. 93). Next, he considers one kind of symmetry, defined by a group of extensions, the real embodiment of which in nature is the shell Turritella duplicata.

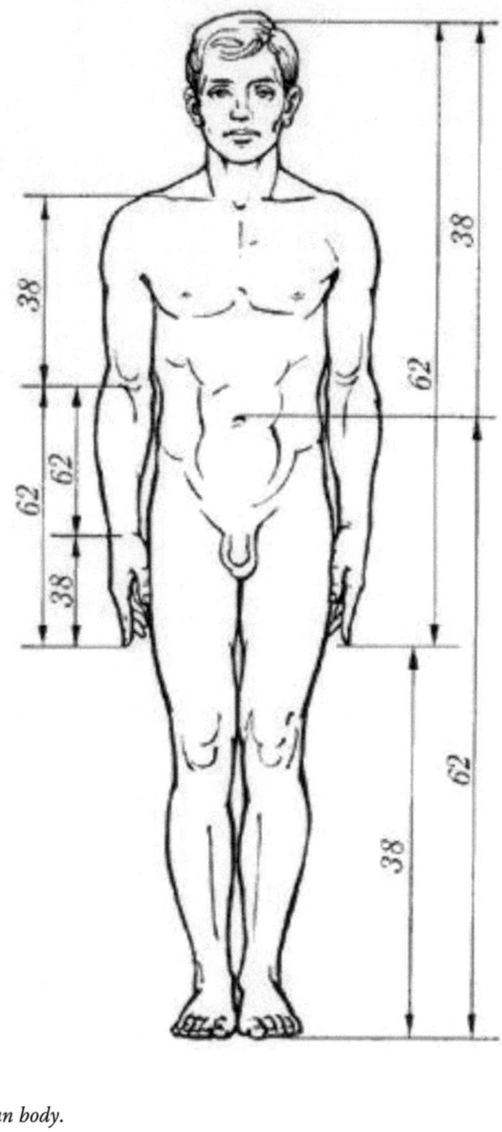

Figure 3.
Golden ratio in the human body.

Special attention from the point of view of the laws of beauty in addition to symmetry attracts the "Golden ratio." The ratio of the Golden ratio is now often used in a variety of spheres of life. But the history of this concept goes back to ancient times when such Sciences as mathematics and philosophy were just emerging. As a scientific concept, the Golden ratio came into use in the time of Pythagoras, namely in the sixth century BC. But even before the knowledge of such a ratio, in practice, it was used in Ancient Egypt and Babylon. A striking evidence of this is the pyramids, for the construction of which used just such a Golden proportion. Apparently the term "Golden ratio" was introduced by Leonardo da Vinci. This term denotes the division of a segment, in which one part of it is as many times larger than the other, as many times smaller than the whole segment. If you make the necessary calculations, you can find the ratio of the greater part of the segment to the smaller. This constant number in the middle ages was called the divine proportion, and is now in our day called the Golden ratio, the Golden mean, or the Golden proportion. This number is usually denoted by the Greek letter Φ; it is approximately equal to 1,61803 … . Roughly speaking, the most part of the segment in this division is 62%, and the smaller 38% of the length of the entire segment.

The Golden ratio is a truly incredible concept, and throughout history, we can find many interesting facts about this proportion. Over time, the Golden ratio rule became an academic routine, but the German scientist Adolf Zeising in 1855 gave it a second life. He published his work entitled "Aesthetic studies." In his work, he presented the Golden ratio as an absolute concept that is universal for all phenomena both in nature and in art. A. Zeising was able to prove that the Golden ratio; in fact, it is the average law for the human body. From the study of A. Zeising, it follows that the main indicator of the Golden ratio is the division of the body by the navel point. And the male body is a little closer to the Golden ratio than the female. Also, the Golden proportion can be observed in other parts of the body, such as, for example, the hand (**Figure 3**).

The Golden ratio is used in painting, sculpture, in the construction of temples, and is found in music and poetry.

3. Methodology and research methods

The study used the following methods: analysis of pedagogical and methodological literature, comparative, historical, and logical types of analysis of the problem of beauty in nature, culture, and science. The methodological basis of the study is a post-non-classical methodology based on synergetic worldview. In post-non-classical science, Trinitarian methodology has also been increasingly used. This methodology assumes the presence of a third element besides two binary oppositions, which is necessary to solve the problem of contradiction of binary oppositions, their integration into a single whole, as a condition of their coexistence. In a sense, a generalization of Trinitarian methodology is a new direction in philosophy-neutrosophy, created recently by Florentin Smarandache, which instead of one third element considers a whole set of neutral elements [9]. The advantage of this philosophical direction is its reliance on a number of new mathematical theories, which are based on non-standard analysis, created by Abraham Robinson [10].

4. Main results

The end of the twentieth century brought a new understanding of beauty in the universe. The synergetic worldview that emerged during this period allows us to give a new aesthetic assessment of the creative role of chaos, its ability under certain conditions to self-organization. Self-organization is characteristic of many complex systems. It consists in the fact that very often a large or even infinite number of quantities or variables characterizing an object "obey" only a few variables, the so-called parameters of order. All these processes are described in synergetics. Traditional science was rejecting the existence of a certain role of chaos in the process of knowledge, referring it to disorganizing factors. However, its constructive role is becoming increasingly apparent nowadays.

Chaos throughout the history of world culture had a negative connotation, and the harmony of the universe was understood solely as overcoming the original chaos with the help of order. In the synergetic worldview, chaos appears as a mechanism of access to attractor structures. To fight against chaos is pointless, because the presence of chaos is a hallmark of complex open systems; you just need to learn how to use its constructive role.

Beauty, since the time of Socrates, in accordance with the prevailing views for centuries, was understood only as a stable order and symmetry. In particular, Socrates argued that beauty is expediency. However, some other philosophers of the past held a different point of view; they saw the product of free thought in beauty. In particular, I. Kant believed that beauty is expediency without purpose; it expresses the ability of man to think of nature according to the laws of freedom. Using modern terminology, this idea can be reformulated as follows: beauty is an attractor, the result of the self-organization of nature, or the flight of free human thought. Synergetic paradigm opened a new vision of beauty as the interaction of space and chaos, their harmonic balance.

As Y.V. Tabakova and A.V. Voloshinov noted: "the Cosmos is beauty relevant, whereas chaos is beauty potential. Again, it is easy to see how the properties of actual and potential beauty in the philosophy of beauty—aesthetics—echo the properties of actual and potential infinity in the philosophy of mathematics, which drew attention to Aristotle" [11].

As it was mentioned above, one of the most common types of symmetry is the similarity transformation. But the similarity transformation is also at the heart of another concept now widely used in mathematics—the concept of fractality. This most important concept underlying beauty and harmony arose at the end of the twentieth century in connection with the development of synergetic worldview, chaos theory, as well as computer technology. Today, this concept as well as the concepts of fractal geometry and fractal graphics have become common among mathematicians and computer artists.

The word fractal was proposed in 1975 by an American mathematician Benoît Mandelbrot. He identified with this word the structures to which his research was devoted. There are several definitions of fractality. One of them is based on such an important property of the fractal as self-similarity. An object is called self-similar when the enlarged parts of the object resemble the object itself and each other. Moreover, self-similarity is understood not only in the classical sense, when the part is an exact copy of the whole, but also in the nonclassical nonlinear sense, when the part is "similar" to the whole. Paraphrasing this definition, we can say that in the simplest case, a small part of the fractal is a reduced copy of the whole (at least approximately), i.e., it contains information about the whole fractal. For example, a snowflake carries information about a snowdrift, and a rock has the same shape as a mountain range.

Theoretically, the self-similarity of the fractal parts is infinite, but visually the human eye is able to distinguish no more than 5–7 fractal self-similarity. The greater number of self-similar parts can be distinguished only by computer magnification.

In the unexplored elements of such structures, the whole mysterious world of the universe can be represented.

The first fractal sets appeared in mathematics long before the works of B Mandelbrot, in the late nineteenth—early twentieth century, but fractals originally, in contrast to symmetry, caused much dislike and bewilderment of many mathematicians of the time. One of the founders of set theory, Georg Cantor, was the first to construct a fractal set from a segment by dividing the segment and throwing out an infinite number of intervals of different lengths from this segment. The result is a fractal object—Cantor dust (**Figure 4**).

Later, other fractal sets were constructed (the Serpinski triangle, the Koch snowflake, etc.), which attracted attention for their aesthetic appeal (**Figure 5**).

Thanks to the works of Mandelbrot [12] in addition to the beauty of symmetry, the aesthetic natural beauty of fractals was discovered. This phenomenon is closely related to the laws of beauty and is often found in living and inanimate nature, as well as in art, architecture, and other spheres of human activity. Fractals give rise to really colorful, original paintings, not inferior to the works of abstract painting (**Figure 6**).

Fractal sets contribute to the emergence of a new look at the aesthetic appeal of mathematics and contribute to the creation of the human ability to "see" the mathematical in the nonmathematical, which is the defining role of fractal geometry for

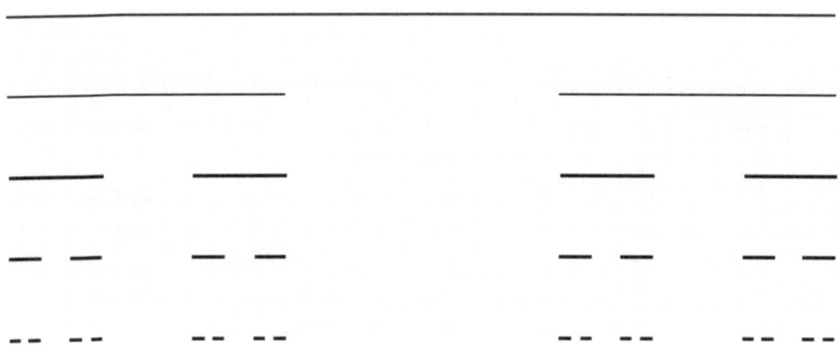

Figure 4.
Cantor Dust.

Figure 5.
Sierpinski Triangle.

Figure 6.
The beauty of fractals.

the future of science and education. In fact, with the advent of fractals, a new kind of computer graphics is born—fractal graphics. With its help, you can create a planar set and the surface of a very complex shape.

Fractal graphics attract our eyes, and software tools for its creation can be the step that will allow us to get closer to this fractal creativity. Familiarity with the elements of fractal geometry contributes to the formation and development of human creativity and artistic (aesthetic) component of his personality.

The idea of "self-similarity" expresses the fact that the hierarchical principle of the organization of fractal structures does not undergo significant changes when viewed through a microscope with different magnification. As a result, these structures look on average the same on small scales as they do on large ones. B. Mandelbrot on this occasion notes that the word "similar" does not always have the classical meaning of "linearly increased or decreased", but is always in agreement with a convenient and broad interpretation of the word "similar". Self-similarity is one of the fundamental principles of the organization of the structure of the world. Information presented in the form of a visual self-similar model is quite simply perceived, reflecting the essential aspects of the object of study in self-similarity and, therefore, visual for the subject working with the model.

Therefore, in creating artificial visual self-similar models, we essentially use one of the fundamental properties of the geometry of nature to organize the perception of the knowledge we are interested in and then we are perceived by this knowledge through the usual visual structure that surrounds us everywhere. In fact, there are more self-similar objects in the real world than we imagine. These include trees and the human circulatory system, galaxies, and snowflakes and these are just the simplest examples.

Under the definition of a fractal, as a structure with self-similar parts, not only designs created with the help of computer graphics are suitable, but also long-known children's pyramids, matryoshkas, and works of art, in which there is a uniformity of repetitions, which is especially the characteristic of folk tales and songs.

In mathematics, fractal (self-similar) structures are not only in geometry. Algebra and number theory also have examples of self-similar structures.

Self-similarity is clearly seen in the very first numerical system with which mankind was acquainted in its historical development—natural numbers. The first natural number 1 is represented by one dash |, the number 2-by two dashes ||, the number 3-by three dashes |||, etc.

In other cases, self-similarity is a little more difficult to see. For example, in decimal fractions 2,14514545145454 51…, the principle of self-similarity in the sequence of digits is as follows: after the first unit should be a combination of digits "45," after the second unit "4545", after the third "454545", etc. In the recording of a given fraction, you can find a cut of numbers of the form "4545… 45" of any finite length which reflected an even number of digits; moreover, none of these segments of the digits does not contain a unit. It is not difficult to see that this decimal fraction is infinite nonperiodic and represents an irrational number.

For algebraic problems with self-similarity of expressions, there are usually simple and beautiful solutions.

For example, calculate the product:

$$\left(\sqrt{2}-1\right)\left(\sqrt{2}+1\right)\left(\sqrt{3}-\sqrt{2}\right)\left(\sqrt{3}+\sqrt{2}\right)\cdot\ \dots\ \cdot\left(\sqrt{99}-\sqrt{98}\right)\left(\sqrt{99}+\sqrt{98}\right)$$

In this example, self-similarity is manifested in the fact that the product consists of pairs of multipliers of the form $\left(\sqrt{n+1}-\sqrt{n}\right)\left(\sqrt{n+1}+\sqrt{n}\right)$, which are arranged in the usual order of natural numbers. In solving this problem, it is important to see that the product of such a pair of factors is 1. Then, distributing the factors in pairs: $(\sqrt{2}-1)(\sqrt{2}+1)=1$, $(\sqrt{3}-\sqrt{2})(\sqrt{3}+\sqrt{2})=1$, …, we find that the desired product is 1.

Self-similar structures in algebra and number theory can be represented by beautiful multi-story radicals and chain fractions.

$$\sqrt{a+\sqrt{a+\sqrt{a+\sqrt{a+\cdots}}}}$$

$$a+\cfrac{1}{a+\cfrac{1}{a+\cfrac{1}{a+\cdots}}}$$

The most famous of the numerical self-similar structures is the Fibonacci numbers' sequence structure. These numbers were discovered by an Italian mathematician of the middle ages Leonardo of Pisa, better known as Fibonacci. After his discovery, these numbers began to be called in the name of a famous mathematician. The amazing essence of the Fibonacci number sequence is that each number in this sequence, starting with the third, is obtained from the sum of the previous two numbers. These numbers form an infinite sequence:

$$0, 1, 1, 2, 3, 5, 8, 13, 21, 34, 55, 89, 144, 233, 377, 610, 987, 1597, 2584,\ \dots$$

which is called the Fibonacci sequence.

There is one very interesting feature in Fibonacci numbers. When dividing any number from a sequence by the number in front of it in a row, the result will always be a value that fluctuates around the already familiar number Φ from the Golden section Φ = 1.61803398875… and through times the a bit more, then a bit less his. But for sufficiently distant two adjacent numbers in the sequence, this result of division becomes almost constant and equal to the number Φ.

It turns out that this remarkable number Φ can be represented as two different self-similar structures.

$$\Phi = \sqrt{1+\sqrt{1+\sqrt{1+\cdots}}} \quad (1)$$

$$\Phi = 1+\frac{1}{1+\frac{1}{1+\frac{1}{1+\cdots}}} \quad (2)$$

There are examples of isomorphic fractal structures based on a number of Golden sections, which allow us to talk about a single fractal grammar of art and attribute the property of fractality to the metalanguage of different arts in nature. Moreover, according to A.V. Voloshinov, there is every reason to say that "the book of nature is written in the language of fractals"; in addition, there are attempts to prove that the book of art is written in the language of fractals.

As we noted above, the transformation of similarity, in particular self-similarity, is a special case of symmetry; then, on the one hand, fractality can be considered one of the manifestations of symmetry. A.V. Voloshinov adheres to an equally wide understanding of symmetry: "since symmetry today is understood expansively as the preservation (invariance) of a certain characteristic, then the varieties of symmetry should include proportion as an invariant of growth, and the Golden ratio as a geometric proportion that has an additive property, and rhythm as a portable symmetry in space or time, and, finally, fractals as self-similar structures that have invariant morphology at different scales" [13].

On the other hand, almost all the types of symmetry we have identified can be considered as special cases of similarity or a combination of similarities, that is, we can consider symmetry as a manifestation of fractality with a finite number of iterations. Thus, the concepts of symmetry and fractality are closely interrelated.

As time goes on, it becomes increasingly clear that the relentless interest in fractals is due not so much to a peculiar fashion and novelty, as to the new opportunities that are opened up to modern Sciences thanks to fractality.

According to recent physical representations, the universe consists of an infinite number of nested fractal levels of matter with similar characteristics. Fractality, according to some philosophers, is one of the universal fundamental properties of being. With the advent of fractals, the limitation of the description of nature with the help of Euclidean geometry was clearly manifested. The world around us is much more diverse than the classical description; it turned out to be a lot of objects described using fractals.

Fractal geometry is not only a new direction in mathematics. Fractal theory is used in geology, geochemistry, hydrodynamics, oceanology, biology, and hydrology. Fractal sets have found application in animated films. But perhaps the main application of fractals is modern computer graphics.

Fractals as well as symmetry have aesthetic appeal; it does not require additional knowledge and skills to feel their natural beauty, to experience aesthetic pleasure from this beauty. It is also worth getting acquainted with fractals in order to understand the beauty of chaos, to learn a new nonlinear world, and awareness of the process of scientific knowledge of the world is one of the most important qualities of a cultural person. Therefore, this new direction in mathematics has great methodological, developmental, and applied potential and requires its gradual introduction, both in the university and in the school curriculum in mathematics [14, 15].

Trinitarian methodology and neutrosophy play an important role in understanding the role of beauty in the universe. From the standpoint of these methodologies, symmetry and fractality are two opposites in the beauty of the universe, mutually complementing each other, aesthetically and mathematically mutually passing into each other. Symmetry reveals in beauty a stable order, and fractality reflects in beauty the result of the self-organization of the chaos of nature or the freedom of human thought.

5. Conclusions

Since the time of the ancient Greeks, space and chaos have been considered oppositions in the concept of beauty. Beauty is determined by the interaction, the harmonic balance of these two oppositions. Symmetry embodies the cosmos, which reveals in beauty a stable order, and fractality penetrates into the second opposition, which reflects in beauty the result of the self-organization of the chaos of nature or the freedom of human thought. Thus, symmetry and fractality are two sides of beauty, aesthetically complementing each other. The synergetic worldview opens to humanity a new vision of beauty—beauty as a synthesis of symmetry and fractality, the most important components of the scientific picture of the world. The interrelated study of symmetry and fractal geometry contributes both to the increase of interest in the study of mathematics and aesthetic education.

Author details

Vladimir A. Testov
Mathematics Department, Vologda State University, Vologda, Russia

*Address all correspondence to: vladafan@inbox.ru

References

[1] Voloshinov AV. Mathematics and Art: 2nd Edition, Revised and Supplemented. Moscow: Prosveshchenie; 2000

[2] Varga B, Yu D, Lazaris E. Language, Music, Mathematics. Per s Wenger. Moscow: Mir; 1981

[3] Tarasov LV. This Amazingly Symmetrical World. Moscow: Prosveshchenie; 1988. p. 176

[4] Poincare H. About Science. Moscow: Nauka; 1990

[5] Boltyansky VG. Mathematical culture and aesthetics. Mathematics in School. 1982;**2**:40-43

[6] Birkhof G. Mathematics and Psychology. Moscow: Sov. Radio; 1977

[7] Sarantsev GI. Aesthetic Motivation in Teaching Mathematics. Saransk: PO RAO: Mordovia pedagogical Institute; 2003. p. 136

[8] Weyl H. Symmetry. Nauka. Glavnaya Redaktsiya Fiz-mat. Literatury: Moscow; 1968. p. 192

[9] Smarandache F. Neutrosophy/ neutrosophic probability, set and logic. Rehoboth: American Research Press; 1998

[10] Robinson A. Non-standard Analysis. Princeton, NJ: Princeton University Press; 1996

[11] Tabakova YV, Voloshin AV. Chaos ancient and modern. Man. 2018;**4**:49-65

[12] Mandelbrot B. Fractal Geometry of Nature. Moscow: Institute of Computer Research; 2002. p. 656

[13] Voloshinov AV, Shingle SV. Harmony—symmetry—beauty. Man. 2017;**4**:81-93

[14] Smirnov EI, Sekovanov VS, Mironkin DP. Increase of educational motivation of schoolchildren in the process of mastering the concepts of self-similar and fractal sets on the basis of the funding principle. Yaroslavl Pedagogical Bulletin. 2015;**3**:37-42

[15] Testov VA. Beauty in mathematical education: Synergetic worldview. Education and Science. 2019;**21**(2):9-26. DOI: 10.17853/1994-5639-2019-2-9-26

Chapter 5

Cosmetovigilance in Hair Straighteners: Determination of Formaldehyde Content by Spectrophotometry and Label Evaluation

Valdicléia Massilon de Abreu,
Maria da Glória Batista de Azevedo
and Juliana de Souza Alencar Falcão

Abstract

The aim of this work was to identify and quantify formaldehyde present in commercial hair straightening formulations, the application of cosmetovigilance from organoleptic/physicochemical tests and label analysis being approached. Samples A1, A3, A5, and A8 had a formaldehyde concentration ranging from 1.5 to 3.83% (w/v), corresponding to concentrations of 7.5, 16.45, 7.9, and 19.15 times higher than that allowed by the National Agency of Sanitary Monitoring (ANVISA), resulting in strong odor characteristic of this active substance. Of these samples, A3 and A5 did not indicate the presence of formaldehyde on the label besides ignoring the warning information and restrictions of use. The absence of the registration number granted by ANVISA for sample A5, which may be an indication of a clandestine product, was also verified. As to the organoleptic and physicochemical properties, only the A2 sample presented different viscosity and centrifugation results. In view of these results, it was concluded that 50% of the analyzed products were reproved because of the presence of formaldehyde out of the allowed concentration, being evident the importance of the implantation of cosmetovigilance system to guarantee the final quality of the cosmetic products, mainly in view of the safety and efficacy of these products.

Keywords: hair straightening, formaldehyde, mandatory labeling

1. Introduction

The beauty market, including toiletries, perfumes, and cosmetics, is one of the fastest growing of all market segments. According to data collected by the Brazilian Association of Personal Hygiene, Perfumes and Cosmetics Industry, Brazil is the third largest sales market with a turnover of approximately R $ 29.4

billion in 2011 and the world leader in hair straighteners and conditioners, with 37.3 and 18.8% of the market, respectively [1]. Hair is increasingly exhibited as a form of expression and affirmation of the personality, and therefore, the cosmetics market presents a range of products with the resources to treat and embellish them [2]. The progressive brush is a hair straightening procedure, introduced in the Brazilian beauty salons in recent years, which contains formaldehyde in its composition and promises a lasting smoothing, around 1–4 months, becoming a fever in beauty salons [3].

According to the National Agency of Sanitary Monitoring (ANVISA), the incorporation of formaldehyde into the hair straightener is prohibited, as it can cause serious damages to the user of the product and to the professional that applies it, such as irritation, pain, and burn in the skin, injuries in the airways, and irreversible damage to the eyes and hair [4]. Health legislation permits the use of formaldehyde in cosmetic hair products only as a preservative in a maximum concentration of 0.2% and as a nail hardener at a concentration of up to 5%, in accordance with Resolution 15 of 2013 [5].

In order to guarantee the safety and efficacy of cosmetic products, the cosmetovigilance system was created and implemented in Brazil, through Resolution RDC No. 332, dated December 1, 2005 [6]. This resolution was elaborated from the MERCOSUR resolution to the member countries to implement this system, being delegated to the competent national bodies of each associated country [7].

Cosmetovigilance has the function of monitoring the response that the product will cause in the market, analyzing the adverse events caused by cosmetics, identifying the risk involved in the use of these products, and taking pertinent behaviors according to the established cause relationship [8, 9].

Despite the risks and prohibitions, there is a variety of products sold to beauty salons that contain formaldehyde in its composition. These products are used for the hair straightening process, and there seems to be an ignorance of the legislation by hairdressers, providing the indiscriminate use of this substance for this purpose [10]. Therefore, the chapter proposes to identify and measure formaldehyde in commercial formulations of permanent hair straighteners and progressive, as well as evaluating the organoleptic and physicochemical properties and correlate results from the information provided by manufacturers on the labeling of the products, in accordance with the annexes IV and V of DRC 211/2005.

2. Material and methods

2.1 Reagents

For the accomplishment of this research, the following reagents were used: formaldehyde PA 37% (Impex), 99% chromotropic acid (disodium dihydrate salt, Sigma-Aldrich), and 98% magnesium sulfate (Sigma-Aldrich), all of analytical grade. Distilled water was used for the preparation of all solutions.

2.2 Equipment used in the experiment

Spectrum spectrophotometer SP 1102; Nova thermostatic shower, model NI 1254; pH meter Hanna, model pH 21; analog rotational viscometer MDJ-1; Centribio centrifuge, model 80-2B; and analytical balance Edutec model FA-2104 N.

2.3 Methods

2.3.1 Collection of samples

For the development of the research, eight samples of permanent and progressive hair straighteners in cosmetic cream from different manufacturers were collected in salons of the municipality of Cuité, PB. The samples were named A1–A8, and their qualitative compositions are described in **Tables 1** and **2**.

2.3.2 Dosing of formaldehyde

The methodology used to identify and quantify the formaldehyde content in progressive brush products was applied according to the standard method

SAMPLES	QUALITATIVE COMPOSITION
A1	Laurileter; Citric acid; Hydroxyestearyl; Berrenyl cetyl; Cetyl alcohol; Frangrance and Aqua (water); Theobrama Cacao; Collagen; EDTA; BHT; Cetitrimethylammonium chloride; Lanolin; Tocopheryl acetate; Magnesium acetate; Formaldehyde; Phenoxyethanol and Isobutyl Paraben.
A2	Aqua, Parfum (amyl cinnamal, anise alcohol, benzyl alcohol, benzyl benzoate, butyphenyl methypropional, citral, citronellol, coumarin, eugenol, geraniol, hexyl cinnamal, limonene, linalool); Cetearyl alcohol; cyclomethicone; Sodium PCA; Cera Alba; Isononyl Isononanoate; Behentrimonium methosulfate; Isosorbide Decaprylate; Aspartic Acid; Arginine; Proline; Tryptophan; Glutamic Acid; Cysteic Acid; Glycine; Leucine; Serine; Butylene Glycol; Cetrimonium Chloride; Argania Spinosa Kernel Oil; Cocos Nucifera Oil; Gardenia Tahitensis Flower; Tocophefol; Euterpe Oleracea Fruit Extract, Bixa Orellana Seed Extract; Paulinia Cupana Fruit Extract; Shea Butter Cetyl Esters; PEG 90M; Glyoxylic Acid; Carbocysteine; Oxalic Acid; Acetamide MEA; (Methylisotiazolinone, Phenethyl Alcohol; PPG-12 Methyl Ether); Cinnamomum Zeylanicum Bark Extract; Corus Calamus Root Extract; Commophora Myrrha Resin Extract; Olea Europaea (olive) Fruit Oil; Citric Acid; Malic Acid; Tartaric Acid; Erythorbic Acid; Gallic Acid; Boric Acid; Sorbic Acid; Oxoacetamide Carbocysteine and Oxoacetamide Amino Acids, Acetic Acid; Benzoic Acid.
A3	Aqua; Cetearyl Alcohol; Cetil Alcohol; Ceteareth-20, Glyceryl Stearate; Behentrimonium Methosulfate/Cetearyl Alcohol, Cetrimonium Chloride; Quaternium-70; Cyclomethicone; Propylene Glycol; Poliquaternium-55; Creatine; Hydrolyzed Keratin; Methyparaben; Propylparaben; Disodium EDTA; Citric Acid; Parfum.
A4	Polyquaternium-67; Aqua (Water); Peg-14m; Butyrospermum Parkii (Shea Butter); Behentrimonium Methosulfate; Cetearyl Alcohol; Behentrimonium Chloride; Isopropyl Mirystate; Glyoxyloyl Carbocysteine and Glyoxyloyl Keratin Amino Acids; Aminopropyl Phenyl trimethicone; Methychloroisothiazolinone; Methylisothiazolinone; Parfum (Fragrance).

Table 1.
Qualitative compositions described on the labels of the capillary straighteners (A1–A4).

SAMPLES	QUALITATIVE COMPOSITION
A5	Água Hidrolized Keratin PG-40; Hidrogenated Castor oil; Cetrimonim Chloride; Pantenol; Laureth-23; Glicerin; Propilene Glicol; Hidrolized Wheatprotein; Poliquaternim-10; Hidroxiethilcellulose PG-40; Whale spermaceti with essence (active principle).
A6	Aqua/Water; Propylene Glycol; Cetearyl Alcohol; Ceteareth-20; Etidronic Acid and Phosphoric Acid; Ethanolamine; Thioglycolic Acid; Honey Extract and Propolis Extract and Pollen Extract and Royal Jelly Extract; Sodium Laureth Sulfate; Parfum/Fragrance.
A7	Aqua, Paraffinum Liquidum, Ammonium Hydroxide, Thioglycolic Acid, Cetearyl Alcohol, Avena Sativa, Petrolatum, Propylene Glycol, Polyquaternium-7, Lanolin, Ceteareth-20, Parfum, Etidronic Acid, Amodimethicone (and) Tallowtrimonium Cholride (and) Nonoynol-10.
A8	Aqua; Cetearyl Alcohol; Cetil Alcohol; Petrolatum; Paraffinum Liquidum (Mineral Oil); Butyrospermum Parkii Butter; BHT; Cetyl Lactate; Chenopodium Quinoa Extract; Polyquaternium-10; Cetrimonium Chloride; Citric acid; Hidrolyzed Keratin; EDTA; Dimethicone; Laureth-4; Laureth-23; Formaldehyde; Methylchloroisothiazolinone; Methylisothiazolinone; Parfum; Hexyl Cinnamal; Alphaisomethyl Ionone; Linalol; Coumarin.

Table 2.
Qualitative compositions described on the labels of the capillary straighteners (A5–A8).

recommended by the National Institute of Occupational Safety and Health [11], followed by the modifications described by Gasparini et al. [12].

2.3.3 Preparation of solutions

The stock solution of formaldehyde with of 1000 mg. L^{-1} was prepared using 2.7 ml of 37% (v/v) formaldehyde solution diluted with distilled water in a 1000 ml volumetric flask. An aqueous solution of chromotropic acid (CA) 5% (w/v) was prepared, dissolving 1.25 g of solute with distilled water in a 25 ml volumetric flask. An aqueous solution of magnesium ($MgSO_4 7.H_2O$) 60% (w/v) was prepared, dissolving 60 g of solute with distilled water in a volumetric flask 100 ml.

2.3.4 Analytical curve for the identification of formaldehyde

The identification of formaldehyde in the solutions was done through the reaction that occurs between the chromotropic acid and formaldehyde in the presence of magnesium after heating, thus producing a colored compound, indicating the presence of formaldehyde in a solution. For the determination of the analytical curve of formaldehyde, the following test was performed: 90, 120, 150, 180, 210, 220, and 230 μl of stock solution containing formaldehyde were transferred to test tubes. In Then, 290 μl of solution of 5% (w/v) chromotropic acid and 3.0 ml of solution of 60% (w/v) magnesium sulfate, with stirring. The tubes were heated for 60 minutes in a steam (100°C), followed by cooling to 25°C. The solutions were transferred to 25 ml flasks and the volume filled with distilled water, obtaining the concentrations of formaldehyde (ppm): 3.6, 4.8, 6.0, 7.2, 8.4, 8.8, and 9.2, respectively. The measurements of absorbance were recorded at 535 nm. All reviews were carried out in triplicate, in order to guarantee the accuracy of the results obtained.

2.3.5 Determination of formaldehyde in the sample

For each commercial sample, 3.0 g was weighed and then dissolved in about 20 ml of distilled water, and the final volume was completed to 100 ml, yielding a solution with concentration (C1) of 30,000 μg ml^{-1}. After 5.0 ml of this solution (C1) were diluted in 25 ml of distilled water (C2 = 6000 μg ml^{-1}). Aliquots of 1.0 ml of the solutions (C2) were transferred to test tubes together with 290 μl of chromotropic acid 5% (w/v) and 3.00 ml of magnesium sulfate 60% (w/v). The tubes were heated for 60 minutes in a steam bath (100°C), followed by cooling to 25°C. The solutions were transferred to 25 ml volumetric flasks, and then the volume was filled with distilled water resulting in a concentration (C3) of 240 μg ml^{-1}. The measurements of absorbance were performed at 535 nm.

2.3.6 Organoleptic and physicochemical properties

The organoleptic and physicochemical tests were performed to assess the characteristics of the products in study. The organoleptic and physicochemical characteristics evaluated were odor, color, appearance, pH, viscosity, density, and centrifugation [13], which were correlated between samples.

2.3.7 Determination of pH

For the determination of the pH, the potentiometric method was used. The samples were diluted to 10% (w/v) in distilled water, at room temperature, in triplicate [13–15].

2.3.8 Determination of viscosity

The viscosity was measured in triplicate in an analog rotary viscometer using 40 g of sample, spindle 4 and speed of 6 rpm. Next, the rotor was inserted vertically into the sample free of blister up to the groove of the rotor rod, and the apparatus was leveled, by reading the viscosity according to the operating procedure of the device [13].

2.3.9 Determination of density

The density was determined by pycnometry. Initially the empty pycnometer (M0) was weighed, first, with distilled water (M1) and, finally, (clean and dry) with the sample (M2). The masses were noted for calculation using the following formula [13]:

$$d = \frac{M_2 - M_0}{M_1 - M_0}$$

where d is the sample density in g/cm^3; M0 is the mass of the empty pycnometer, in grams; M1 is the mass of the pycnometer with distilled water, in grams; and M2 is the mass of the pycnometer with the sample, in grams.

2.3.10 Centrifuge test

From each sample, 5 g was placed in centrifugal tubes and submitted to cycles of 1000 and 2500 rpm for 15 minutes each. The homogeneity of the study formulations was evaluated by observing macroscopic separation of phases after performing the described procedure [13, 14, 16].

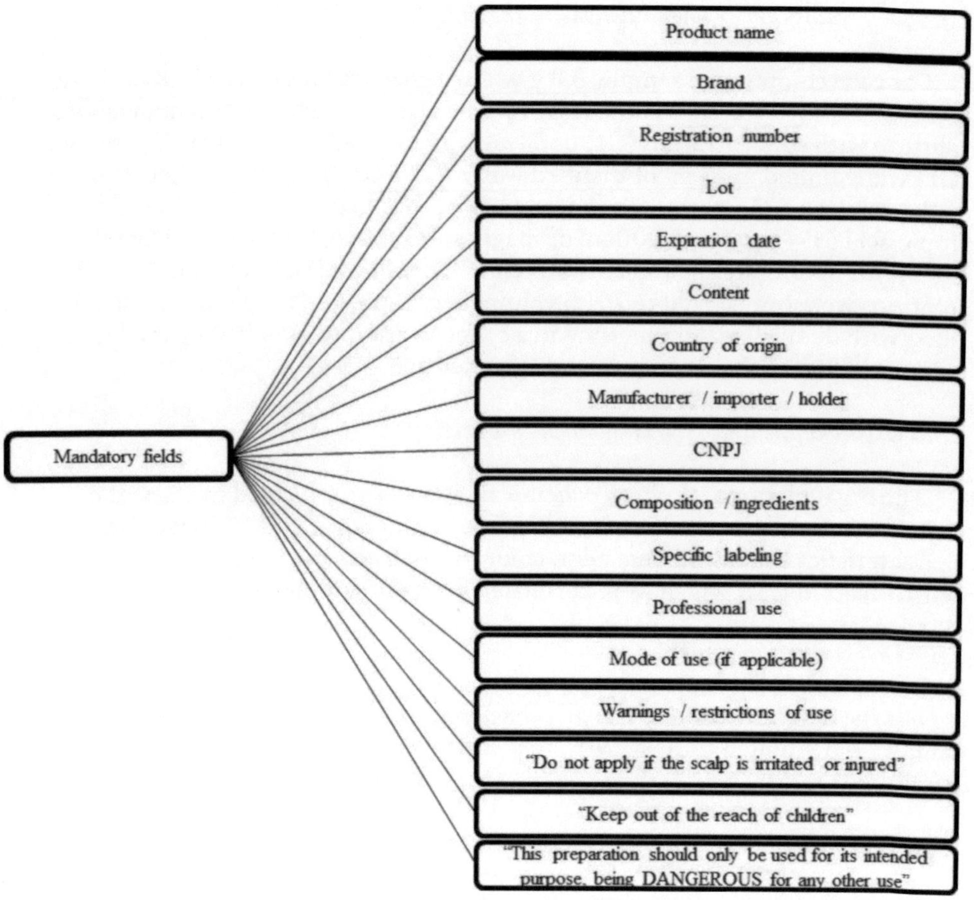

Figure 1.
Mandatory items on labeling of capillaries straighteners according to Annexes IV and V of DRC 211/2005.

2.3.11 Label analysis

A qualitative analysis of the primary and secondary packaging of the tested products by a critical visual investigation was carried out, following the criteria laid down by national legislation—RDC 211/2005 which defines the labeling rules for cosmetic products [17] and RDC 332/2005 which deals with regulation and implementation of cosmetology in the cosmetic industries [6]. The analyzed items were specified according to **Figure 1**.

3. Results

3.1 Analytical curve for the identification and determination of formaldehyde

The identification of formaldehyde was evidenced by the production of a pink compound resulting from the reaction between chromotropic acid and formaldehyde in the presence of magnesium sulfate. The analytical curve data, resulting from the average of three calibration curves, were adjusted by linear regression (**Figure 2**), whose equation of the line is given by absorbance = 0.0922 × [formaldehyde solution, chromotropic acid, and magnesium sulfate] (ppm) − 0.0234. The correlation coefficient obtained was 0.996, a significant linear regression. Samples A1, A3, A5, and A8 developed a coloration after heating, indicating the presence

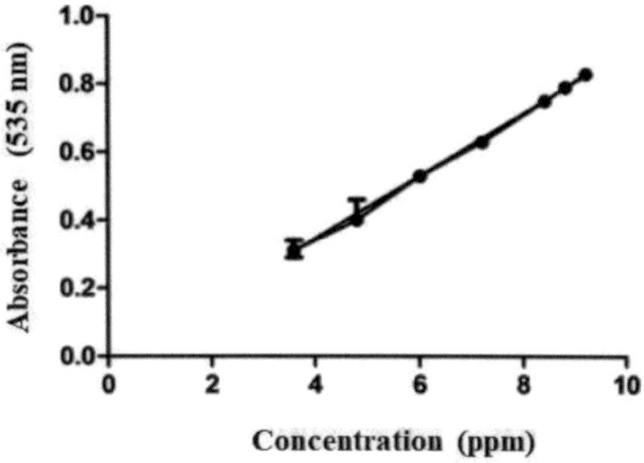

Figure 2.
Graphical representation of the formaldehyde analytical curve using chromotropic acid (AC) and sulfate of magnesium ($MgSO_4$) at 535 nm where Y = 0.0922, x = 0.0234, and R = 0.996.

Sample	Avagerage absorbance	Concentration (ppm) of formaldehyde in diluted sample	Percentage of formaldehyde in formulation (% w/v)
A1	0,31	3,6	1,50
A2	0,00	0.00	0,00
A3	0,71	7,9	3,29
A4	0,00	0,00	0,00
A5	0,33	3,8	1,58
A6	0,00	0,00	0,00
A7	0,00	0,00	0,00
A8	0,83	9,2	3,83

Table 3.
Absorbances, concentrations of formaldehyde in diluted samples (ppm) and percentages of formaldehyde in the formulations A1–A8.

of formaldehyde in these formulations, which tends to be as darker as greater the concentration of said substance. The samples A3 and A8, on the other hand, presented more intense in relation to others, which justifies a higher concentration of formaldehyde. The percentages of formaldehyde calculated by the equation from the analytical curve confirm these results and are described in **Table 3**.

3.2 Organoleptic and physicochemical characteristics

The results obtained in organoleptic and physicochemical tests of the study samples are listed in **Table 4**. Regarding appearance and color, observed macroscopically, the samples presented as homogeneous creams, whether or not colored and without precipitation or exudation. The odor was checked directly through the smell, being possible to smell characteristic of formaldehyde in samples A1, A3, A5, and A8, capable of causing some mucosal irritation and burning in the eyes during observation, confirming symptoms caused by exposure to formaldehyde. The sample A5, although with a strong chocolate odor, failed to mask the presence of formaldehyde in the product. Samples A6 and A7 showed an odor of sulfur, a characteristic of thioglycolic acid. The samples A2 and A4 presented no strong or unpleasant smell, only the essence odor used in these products.

Samples A2 and A4 showed a pH of very acid (1.0 and 1.3, respectively), which may damage the capillary wires [18]. In the other hand, the sample A8 has a pH

ESSAYS	SAMPLES							
	A1	**A2**	**A3**	**A4**	**A5**	**A6**	**A7**	**A8**
Aspect	CH	CH	CH	CH	CH	CH	CH	CH
Color	Salmon	Lilac	White	White	Brown	White	White	White
Odor	F	NO	F	NO	F	T	T	F
pH	3,8 ± 0,04	1,0 ± 0,01	3,6 ± 0,03	1,3 ± 0,01	3,8 ± 0,04	9,2 ± 0,09	9,4 ± 0,09	4,9 ± 0,05
Density g/cm^2	1,0 ± 0,01	0,9 ± 0,009	1,0 ± 0,01	1,0 ± 0,01	0,9 ± 0,009	1,0 ± 0,01	1,0 ± 0,01	1,0 ± 0,01
Centrifugation	N	M	N	N	N	N	N	N

Caption: CH - Homogeneous Cream; F - Strong (formaldehyde); T - Sulfur (thioglycolic acid); NO - No characteristic odor (formaldehyde and sulfur); M - Modified (phase separation); N - Normal (without phase separation).

Table 4.
Organoleptic and physico-chemical characteristics of the formulations under study.

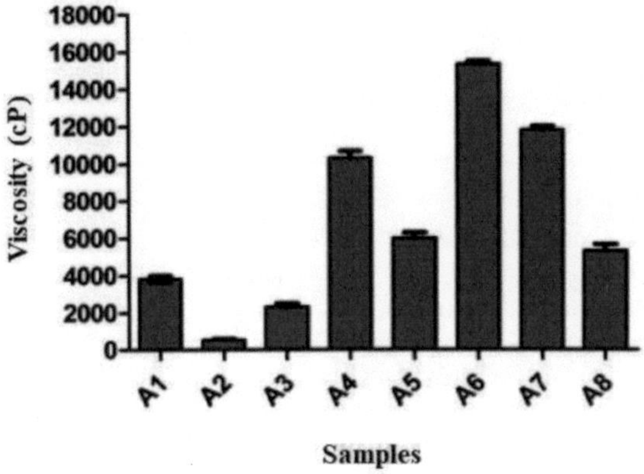

Figure 3.
Average values of the viscosity (cP) of the samples A1–A8 at 25°C.

within the range that tends to assist in the maintenance of wires. The samples A1, A3, and A5 also have a pH slightly acidic, being able to close the capillary cuticle and help in preserving the color deposited in the hair. The samples A6 and A7 presented extremely high pH, a characteristic of products for permanent straightening.

The analyzed samples did not present significant variation in relation to the determination of the density and did not need to correct the weighing to carry out the dosing of the formaldehyde.

The formulations showed without phase separation, precipitation, formation of caking, and coalescence, except the A2 sample that presented phase separation after being subjected to the centrifugation.

The viscosity results can be in **Figure 3**. These results point to discrepant values between formulations, ranging from extremely low (A2) to extremely high (A6 and A7).

3.3 Label analysis

The results of the analysis of the product labels are presented in **Table 5**. Samples A1, A3, A5, and A8 presented formaldehyde content in the procedure

SAMPLES / EVALUATION OF LABEL	A1	A2	A3	A4	A5	A6	A7	A8
Product name	S	S	S	S	S	S	S	S
Brand	S	S	S	S	S	S	S	S
Registration number	S	S	S	S	N	S	S	S
Lot	S	S	S	S	S	S	S	S
Expiration date	S	S	S	S	S	S	S	S
Country of origin	S	S	S	S	S	S	S	S
Manufacturer / importer / holder	S	S	S	S	S	S	S	S
CNPJ	S	S	S	S	S	S	S	S
Composition / ingredients	Fp	Fa	Fa	Fa	Fa	Fa	Fa	Fp
Professional use	S	S	S	S	S	S	S	S
Mode of use (if applicable)	S	S	S	S	S	S	N	N
Warnings / restrictions of use	A	B	B	B	B	B	B	A
"Do not apply if the scalp is irritated or injured"	S	N	N	N	N	S	S	S
"Keep out of the reach of children"	S	S	N	S	S	S	S	S
"This preparation should only be used for its intended purpose, being DANGEROUS for any other use"	N	N	N	N	N	N	N	S

Caption: A - Indicated formaldehyde concentration; B - Concentration of formaldehyde not indicated; N - Not indicated on the label; S - Indicated on label; Fp-formaldehyde present; Fa - formaldehyde absent.

Table 5.
Analysis of the labels of the products under study according to Annexes IV and V to DRC 211/2005.

spectrophotometric assay; however, only samples A1 and A8 indicated formaldehyde in their composition. Samples A3 and A5, in addition to the absence of the substance, presented other irregularities on the label, both ignoring the warning information and usage restrictions. Sample A5 does not yet have registration number granted by ANVISA, evidencing the likely clandestine origin of this product.

The samples A2, A4, A6, and A7, whose labels said formaldehyde-free, are not really content of the substance, being in compliance and complying with the technical requirement legislations.

4. Discussion

According to the NIOSH 3500 [11] reference method, formaldehyde is determined by spectrophotometry through reaction with chromotropic acid in the presence of concentrated sulfuric acid, after heating, to obtain a soluble polymer violet-red that can be detected in length of wavelength of 580 nm. However, although the method is selective, not suffering interference from other aldehydes, it has some drawbacks. The main one is the use of concentrated sulfuric acid by the toxicity and corrosivity that it presents. Therefore, some authors have proposed modification methodology that could minimize these disadvantages [12, 19–23].

According to Toutianoush et al. [22], it is possible to dose the formaldehyde by spectrophotometry in the presence of a solution of magnesium sulfate ($MgSO_4•7H_2O$) instead of sulfuric acid. A red/pink is produced after heating of formaldehyde with an excess of chromotropic acid in a steam bath. In the absence of magnesium sulfate, this reaction does not occur. It is likely that the oxygen atoms of the cyclotetrachromotropylene-hydroxyl are pre-arranged for complexation with magnesium. Gasparini et al. [12] used this method to dose formaldehyde disinfectants and hair care products, considering the adequate method, low operating cost, and simplicity and selectivity, meeting the requirements of green analytical chemistry.

Thus, the applicability of the proposed method to determination of formaldehyde in straightening hair products of different brands proved to be appropriate, possible to identify the presence of formaldehyde in samples A1, A3, A5, and A8 in concentrations ranging from between 1.5 and 3.83% (w/v), i.e., concentrations 7.5; 16.45; 7.9 and 19.15 times that allowed by ANVISA [5].

Some authors also dosed formaldehyde in commercial formulations of hair straighteners using methods such as HPLC, gas chromatography, and mass spectrometry, finding concentrations of 1.6–11.5%, similar results to that of this research, with concentrations above that accepted, evidencing the use of formaldehyde as a hair straightener and not just as a preservative at the maximum concentration of 0.2%, as recommended by health legislation. This shows the lack of supervision by regulatory institutions and the deep lack of knowledge or lack of concern for part of consumers and professional hairdressers in face of the risks that these products bring to health [24, 25].

In addition to the assay, it is also important to analyze the organoleptic and physicochemical characteristics of a cosmetic formulation, because usually the texture, odor, color, and appearance of the product are examined before your purchase; these being important parameters from the commercial point of view, consumers may or may not be attracted to these characteristics. The sample A5 presents a dark-brown tone and chocolate essence; these characteristics tend to seduce the consumer. The color of a product, however, has no connection with the presence or absence of formaldehyde, differently from the intense odor, which may be indicative, since even using extremely high essences fragrant, it is still possible to smell it.

The pH of the hair strands ranges from 4.5 to 5.5. Capillary formulations with extreme pH variations can damage the capillary wire, because the hair shrinks and stiffens or even dissolves completely at very high pH acid or increases porosity as the layers of cuticle dilate, resulting in a dry appearance, opaque, reaching to complete hair dissolution in strongly alkaline pH. The pH values of samples A1, A3, A5, and A8 were compatible with the hair samples, while samples A2, A4, A6, and A7 had a pH out of the tolerated range, the integrity of the capillary wire [18]. In this way, products for professional use should be used with caution and applied by trained professionals.

Viscosity is a measure of resistance of a flow system when submitted to a mechanical stress. Therefore, the higher the viscosity, the higher the resistance and the force to be applied to flow with a certain velocity [26]. In the case of the products tested, there was a considerable variation of this parameter, of samples having viscosities less than 4000 cP to samples with viscosities above of 12,000 cP, thus presenting flow profile applicability. The viscosity results can also inform the influence of the sedimentation rate of the droplets present in the cosmetic cream form, complementing the result of the centrifugation.

The centrifugation test provides quick information about phase separation, thus predicting whether the product will separate as a function of time [13], which can lead to significant differences in the content of assets. In the present study, only the A2 sample presented separation of post-centrifugation phases, which can be influenced by the low viscosity of the sample (566 ± 57.7 cP). According to Stokes's law, the speed of sedimentation is inversely proportional to viscosity and directly proportional to the size of the droplets, the gravity, and the density difference between the dispersed medium and dispersant [27].

The evaluation of product quality parameters for progressive brush becomes essential, once beauty salons are frequently visited by sales promoters that offer products that are capable of providing a "true miracle" in straightening the hair without formaldehyde in its composition. However, they may have a high concentration, masked by an extremely fragrant and unidentified formula of its presence on the label [28], which could be proven in the present study, where 50% of the products for hair straightening analyzed were disapproved due to the presence of formaldehyde out of the concentration allowed, as well as irregularities in the labels.

5. Conclusion

The chapter is important to the cosmetovigilance system to ensure the final quality of cosmetic products, taking into safety, efficacy, and information to ANVISA, to the manufacturer, and to the consumer. In this way, this system may provide numerous benefits to the cosmetics industry as a whole. As measures, it would be important to distribute orientation books and flyers of cosmetic products since the vigilance is not only the responsibility of companies but also of the consumers and competent institutions.

Author details

Valdicléia Massilon de Abreu, Maria da Glória Batista de Azevedo and Juliana de Souza Alencar Falcão*
Universidade Federal de Campina Grande (UFCG), Curso de Farmácia, Unidade Acadêmica de Saúde, Cuité, PB, Brazil

*Address all correspondence to: alencarfalcaojuliana@gmail.com

References

[1] ABIHPEC–Associação Brasileira de Indústria de Higiene Pessoal, Perfumaria e Cosméticos [Internet]. 2011. Disponível em: http://www.abihpec.org.br

[2] Franquilino E. Cabelos através dos tempos. Revista de negócios da Indústria da Beleza. 2009;**4**(11):6-16

[3] Ionta LMP, Silva JO. Efeitos tóxicos do formaldeído em escova progressiva. In: 10ª mostra acadêmica Unimep. Piracicaba: 10° Congresso de Pós-Graduação; 2012

[4] Köhler MEF. A química da estética capilar como temática no ensino de química e na capacitação dos profissionais da beleza [Dissertação]. Santa Maria: Universidade Federal de Santa Maria. Programa de Pós-Graduação em Educação em Ciências: Química da Vida e Saúde; 2011. 113 p

[5] Brasil. Agência Nacional de Vigilância Sanitária. RDC n° 15, de 26 de Março de 2013. Aprova o regulamento técnico e lista as substâncias de uso cosmético: Acetato de chumbo, pirogalol, formaldeído e paraformaldeído e dá outras providências. Brasília: Diário Oficial da União; 27 Mar 2013

[6] Brasil. Agência Nacional de Vigilância Sanitária. RDC n° 332 de 01 de Dezembro de 2005. Dispõe sobre a implementação de um Sistema de Cosmetovigilância, a partir de 31 de dezembro de 2005 em todas as empresas fabricantes e/ou importadoras de produtos de hygiene pessoal, cosméticos e perfumes, instaladas no território nacional. Brasília: Diário Oficial da União; 02 Dez 2005

[7] MERCOSUL. MERCOSUR/GMC/RES N° 19/05. Programa de cosmetovigilância em elarea de productos de higiene personal, cosméticos e y perfumes [Internet]. 2005. Disponível em: http//:www.toxiclin.com.br/DOCUMENTOS/MERCOSUR-GMC-RES.htm

[8] Brasil. Agência Nacional de Vigilância Sanitária. Guia para Avaliação de Segurança de Produtos Cosméticos. 2a edição. Brasília: ANVISA; 2012

[9] Magalhães WCP. Integração do P & D à Cosmetovigilância. Cosmetics & Toiletries. 2012;**24**(2):28-29

[10] Lorenzini S. Efeitos adversos da exposição ao formaldeído em cabeleireiros [Tese]. Porto Alegre: Universidade Federal do Rio Grande do Sul. Pós-Graduação em Ciências Pneumológicas; 2012. 77 p

[11] NIOSH–National Institute for Occupational Safety and Health. Manual of Analytical Methods (NMAN): Method 3500 (2)-Formaldehyde, 4th ed. US Department Health, Education and Welfare; 1994

[12] Gasparini F, Weinert PL, Lima LS, Pezza L, Pezza HR. A simple and green analytical method for the determination of formaldehyde. Journal of the Brazilian Chemical Society. 2008;**19**(8):1531-1537

[13] Brasil. Agência Nacional de Vigilância Sanitária. Guia de controle de qualidade de produtos cosméticos/Uma Abordagem sobre os Ensaios Físicos e Químicos. 2ª edição, revista. Brasília: ANVISA; 2008

[14] Isaac VLB, Cefali LC, Chiari BG, Oliveira CCLG, Salgado HRN, Corrêa MA. Protocolo para ensaios físico-químicos de estabilidade de fitocosméticos. Revista de Ciências Farmacêuticas Básica e Aplicada. 2008;**29**(1):81-96

[15] Brasil. Agência Nacional de Vigilância Sanitária. Farmacopeia Brasileira. 5ª ed. Vol. 1. Brasília: Agência Nacional de Vigilância Sanitária; 2010

[16] Casteli VC, Mendonça CC, Campos MAL, Ferrari M, Machado SRP. Desenvolvimento e estudos de estabilidade preliminar de emulsões O/A contendo Cetoconazol 2.0%. Acta Scientiarum. Health Sciences. 2008;**30**(2):121-128

[17] Brasil. Agência Nacional de Vigilância Sanitária. RDC n° 211 de 14 de julho de 2005. Estabelece a definição e a classificação de produtos de higiene pessoal, cosméticos e perfumes, conforme Anexo I e II desta Resolução e dá outras definições. Brasília: Diário Oficial da União; 18 Jul 2005

[18] Halal J. Tricologia e a química cosmética capilar. EZ2translate. São Paulo: Cengage Learning; 2012

[19] Georghiou PE, Ho CK. The chemistry of the chromotropic acid method for the analysis of formaldehyde. Canadian Journal of Chemistry. 1989;**67**:871-876

[20] Fagnani E, Melios CB, Pezza L, Pezza HR. Desenvolvimento de método espectrofotométrico para análise de paraformaldeído em saneantes comerciais e industriais. Eclet Quím [Internet]. 2002. Vol. 27. Disponível em: http://www.scielo.br/scielo.php?script=sci_arttext&pid=S0100-46702002000100018

[21] Gigante AC, Gotardo MA, Tognolli JO, Pezza L, Pezza HR. Spectrophotometric determination of formaldehyde with chromotropic acid in phosphoric acid medium assisted by microwave oven. Microchemical Journal. 2004;**77**:47-51

[22] Toutianoush A, Schnepf J, ElHashani A, Tieke B. Selective ion transport and complexation in layer-by-layer assemblies of p-sulfonato-calix[n]arenes and cationic polyelectrolytes. Advanced Functional Materials. 2005;**15**(4):700-708

[23] Yasri NG, Seddik H, Mosallb MA. Spectrophotometric determination of formaldehyde based on the telomerization reaction of tryptamine. Arabian Journal of Chemistry. 2011. Disponível em: http://www.sciencedirect.com/science/article/pii/S1878535211000463

[24] Mazzei JL, Figueiredo EV, Veiga LJ, Aiub CA, Guimarães PI, Felzenszwalb I. Mutagenic risks induced by homemade hair straightening creams with high formaldehyde content. Journal of Applied Toxicology. 2010;**30**(1):8-14

[25] Pierce JS, Abelmann A, Spicer LJ, Adams RE, Glynn ME, Neier K, et al. Characterization of formaldehyde exposure resulting from the use of four professional hair straightening products. Journal of Occupational and Environmental Hygiene. 2011;**8**(11):686-699

[26] Almeida IF, Bahia MF. Reologia: Interesse e aplicações na área cosmético-farmacêutica. Cosmetics & Toiletries. 2003;**15**(3):96-100

[27] Allen LV Jr, Popovich NG, Ansel HC. Formas farmacêuticas e sistemas de liberação de fármacos. 8ª. ed. Porto Alegre: Artmed; 2007

[28] Cardoso J, Pires M. Formol em Produtos para Alisamento de Cabelos. Cosmetics & Toiletries. 2011;**23**(4):52-58

Chapter 6

Nanocosmetics: Production, Characterization, and Performance Improvement

Júlia Scherer Santos

Abstract

Nanocosmetics are personal care products containing nanocarriers or nanoparticles. Nanocarriers have been used in sunscreens, moisturizers, perfumes, and anti-aging and hair products. These carriers increase formulation efficacy and promote controlled release of active ingredients. Polymeric nanocarriers or lipid nanocarriers containing sunscreens have enhanced ultraviolet protection. In addition, these nanocarriers protect unstable ingredients from degrading ultraviolet radiations. Furthermore, nanocarriers also impart moisturizing effect due to their reduced particle size. This chapter describes issues related to nanocarrier-based cosmetics production, characterization, and biological evaluation.

Keywords: nanocarriers, polymeric nanocarriers, lipid nanocarriers, nanocosmetics, biological assays

1. Introduction

Nanocosmetics are skin formulations containing nanocarriers or nanoparticles. They have several advantages over conventional cosmetics (ie, cosmetics without nanotechnology). Nanocarriers are nanometric carriers having a substance entrapped inside. They are classified into lipid nanocarriers and polymeric nanocarriers. The first ones include solid lipid nanoparticles, nanostructured lipid carriers, and nanoemulsions. Polymeric nanocarriers include nanocapsules, nanospheres, and micelles. In order to demonstrate the importance of these nanocarriers in cosmetics, an initial examination was performed in Scopus databases and Web of Science database until 2019. The terms "nanocarriers" and "cosmetics" were crossed searching for abstract, article title, and keywords, and the results are shown in **Figure 1** and **Table 1**. The number of publications retrieved from the databases was similar although the Web of Science database recovered a smaller number of publications.

Figure 1 shows the number of publications from search 01 over time. An exponential increase in publications is observed in both databases. It is also noteworthy that the first publications using nanocarriers applied to cosmetics date from the mid-2000s, showing, therefore, that this term is of recent use.

In order to detect the number of publications regarding lipid nanocarriers and polymeric nanocarriers, two new searches were performed. From the search 01, the total number of publications was further crossed with the term "lipid" and "polymer" in search 02 and search 03, respectively (**Table 1**). In this context, the

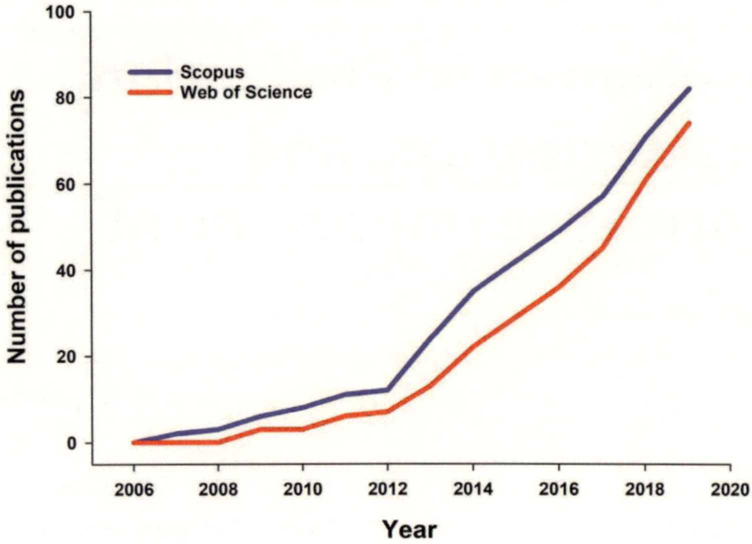

Figure 1.
Total number of publications retrieved in databases until 2019 when the terms "nanocarriers" and "cosmetics" were crossed. Blue line displays publications from Scopus database. Red line displays publications from Web of Science database.

Terms	Database	
	Scopus	Web of Science
Search 01 "Nanocarriers" AND "Cosmetics"	82	74
Search 02 "Nanocarriers" AND "Cosmetics" AND "Lipid"	54	40
Search 03 "Nanocarriers" AND "Cosmetics" AND "Polymer"	16	8

Table 1.
Results from Scopus database and Web of Science database from 2006 to 2019.

vast majority of publications refer to lipid due to its ingredients' biocompatibility. Although polymeric nanocarriers had a smaller number of publications, they are still important, mainly in relation to the use of natural polymers.

Nanoparticles applied to skin formulations comprise inorganic nanoparticles and organic nanoparticles. Inorganic nanoparticles such as titanium oxide nanoparticles and oxide zinc nanoparticles are common in sunscreen formulations. On the other hand, organic nanoparticles have polymers or lipids in their composition. The approach to cosmetics containing inorganic nanoparticles is beyond the scope of this chapter. The main aspects of production and characterization of lipid nanocarriers and polymeric nanocarriers will be addressed in this chapter. Further, *in vitro* and/or *in vivo* evaluation of nanocarriers and nanocosmetics will also be described.

2. Nanocosmetics production and characterization

Nanocosmetics are also called nanocosmeceuticals: cosmeceuticals containing nanotechnology [1]. Cosmeceuticals are cosmetics with benefits over traditional cosmetics. They are defined as quasi pharmaceuticals as they come from the combination of cosmetics and pharmaceuticals. Cosmeceuticals must have an established mechanism of action in skin cells or tissues. These cosmetics are often tested regarding efficacy studies, and in this sense, they resemble pharmaceutical products [1]. Nanocarrier production methods are classified into bottom-up or bottom-down [2] and vary accordingly to each nanoparticle type (ie., matrix types and depot-types) [2, 3]. After nanocarrier production, characterization studies should be conducted. They include the following assays: determination of particle size, zeta potential, encapsulation efficiency, and in vitro release study [2]. Also, characterization studies should be performed on nanocosmetics.

2.1 Nanocarrier production

Nanocarriers are nanometric carriers entrapping the desired substance in its structure. As this substance is enclosed in the nanocarrier, a performance improved could be achieved. There are several methodologies applied to nanocarriers and the present work does not intend to exhaust all existing ones. **Table 2** below shows some of the methods applied to the following nanocarriers: nanoemulsions, solid lipid nanoparticles, nanostructured lipid carriers, nanocapsules, and nanospheres [2, 28].

Polymeric nanocarriers (nanocapsules or nanospheres) are composed of biodegradable and/or biocompatible polymers such as poly (D,L-lactide-co-glycolide) [25] and chitosan [22, 24] and they have a size ranging from 100 to 1000 nm. The substance is encapsulated, dispersed, or adsorbed on the surface of these particles. Nanocapsules are depot-like systems formed by a polymeric wall and an oily core. On the other hand, nanospheres do not contain oil in their composition. Therefore, nanospheres are formed by a polymeric matrix where the substance is dispersed or dissolved [2, 28]. To obtain these particles, methods such as interfacial deposition of pre-formed polymer, emulsification/solvent diffusion, and interfacial polymerization can be employed. Recently, the focus has been on the use of natural polymers [22, 24, 27].

Nanocarrier	Method	References
Nanoemulsions	Ultrasonic emulsification-solvent evaporation	[4, 5]
	Ultrasonication	[6]
	Emulsification/evaporation	[7]
Lipid nanocarriers	High pressure homogenization	[8–13]
	High shear homogenization/high pressure homogenization	[14]
	Melt emulsification coupled with high shear homogenization	[15, 16]
	Double emulsion	[17]
	Hot melt microemulsion	[18]
	Microemulsion coupled with probe sonication	[19]
	Ultrasonication	[6, 20, 21]
Polymeric nanocarriers	Ionic crosslinking	[22]
	Interfacial deposition of pre-formed polymer	[23, 24]
	Polymerization of monomers	[7]
	Emulsification/solvent evaporation	[25, 26]
	Two-step desolvation	[27]

Table 2.
Nanocarrier preparation methods.

Lipid nanocarriers differ from polymeric nanocarriers since they have solid/liquid lipid in their composition. These lipids account for their high biocompatibility. Solid lipid nanoparticles (SLN) are formed by a solid lipid matrix and were developed in the 90s. However, SLN have low long-term stability [28]. Nanostructured lipid carriers (NLC) were developed in order to increase the long-term stability of solid lipid nanoparticles.

Although nanoemulsions are lipid nanocarriers, just as SLN and NLC, they have an internal phase with a nanometric droplet size and different production methods. Therefore, the method to be selected will depend on the type of nanocarrier and also on the equipment and raw materials available.

2.2 Nanocarrier characterization

Nanocarrier characterization is performed to provide information such as particle size, zeta potential, entrapment efficiency [12, 17, 29], morphological aspect [26, 30], and pH [24]. Other analyses also include nanocarrier physical stability [10] and compatibility between ingredients of nanocarriers [31].

Particle size distribution and zeta potential analyses are essential, as they are directly related to nanocarriers' biological behavior. Nanocarriers have different particle size distributions varying according to carrier type [11–13, 22, 24, 25, 30] and also according to their composition [12, 13, 24, 31, 32]. In this sense, lipid nanocarriers' particle size distribution range from 100 nm to 200 nm [12, 13, 20, 29], or they may have particle size distribution ranging from 60 nm to more than 200 nm [32]. For nanoemulsions, droplet size may be smaller than 100 nm [30] or greater than 200 nm [5]. Regarding polymeric nanocarriers, sizes smaller than 100 nm [26], in the range between 100 and 200 nm [22], or even greater than 200 nm [25] are described.

Zeta potential reflects the particle surface charge and is related to its composition. Negative zeta values are the most reported [17, 23, 24]. Nevertheless, positive zeta potential is also reported for nanocarriers since there is an increase in its biological efficacy compared to negatively charged nanocarriers [29]. Entrapment efficiency, on the other hand, is related to the preparation method and to the physicochemical properties of the encapsulated substance [22]. This methodology sets the percentage encapsulated in nanocarriers [5, 15, 25–27, 32]. High encapsulation efficiency can increase biological efficacy [15, 25].

In addition, methodologies such as atomic force microscopy [18, 26], transmission electron microscopy [11, 21, 26, 30], scanning electron microscopy [7, 19, 22], and optical microscopy [8, 19] have been used as way to complement size determination and to evaluate particle morphology. Moreover, pH measurement of nanocarriers [24, 30, 33, 34] allows assessing the compatibility of the skin to the nanocarrier.

Physical stability is another approach used to monitor the formulation behavior by measuring particle size distribution, zeta potential, and entrapment efficiency over a predetermined period of time [8, 10, 12, 22, 34]. Physical stability has also been showed by equipment that detects instability phenomena, such as sedimentation and cremation [24].

Furthermore, ingredient compatibility [14, 25–27, 35] must be also be performed prior to the development/preparation of any nanocarrier using methodologies such as thermal analysis. Ingredient interaction is not desired in many cases. Nevertheless, in some situations, ingredient interaction can be beneficial since it causes increase in cosmetic efficacy [27]. Thermal analyses are also used for other purposes: to evaluate lipid crystallinity, to measure the physical stability of lipid nanoparticle, and to show that substance is encapsulated within the nanoparticle [16, 29]. Finally, in the case of nanoemulsions, other analyses can also be performed: electrical conductivity [30, 33, 34], phase diagram [30], and interfacial tension [33].

Cosmetic form	Main component	References
Cream	Stearic acid/triethanolamine	[19, 26]
	Sodium polyacrylate/dimethicone/cyclopentasiloxane/ trideceth-6/ PEG/PPG-18/18 dimethicone	[23]
	Glyceryl monostearate	[25]
	Hydroxyethyl acrylate/sodium acryloyldimethyl taurate	[27]
Hydrogel	Ammonium acryloyldimethyltaurate/VP copolymer	[23]
	Carbomer	[14, 17, 20]
	Chitosan	[24]
Gel-cream	Cetearyl alcohol/dicetyl phosphate/Ceteth-10 phosphate Acrylates/C10-30 alkyl acrylate crosspolymer	[18]

Table 3.
Cosmetics forms employed in nanocosmetics.

2.3 Nanocosmetics production and characterization

Once properly characterized, nanocarrier can be added in an appropriate semi-solid form since nanocarriers are usually aqueous liquid forms that have low viscosity [28]. Nevertheless, it is also possible to thicken the nanocarrier solution/ dispersion [34]. Cosmetics forms employed in nanocosmetics (**Table 3**) include mainly creams [12, 19, 23, 25], hydrogels [14, 17, 20, 23, 24], or gel-creams [18].

From the moment that nanocosmetics are obtained, they must be characterized by tests such as pH, organoleptic characteristics, and rheology [19, 23–25]. As previously mentioned, pH determination is a prime analysis to ensure formulation compatibility with the skin. In contrast, organoleptic characteristics such as appearance and color [24] can be visually determined. In the case of color determination, it is also possible to monitor it by equipment [20]. Rheological analysis defines the product ease/difficulty of flow. Consequently, rheological behavior is crucial because it is directly related to spreadability of cosmetics in the skin [23].

Stability determination of a nanocosmetic is a common assay comprising analyses such as organoleptic characteristics, pH, rheology, and entrapped substance content [18, 19, 23, 24]. Taking into consideration that stability test submits the product to temperature variations, changes in pH, rheology, and substance content are expected. In that respect, nanoencapsulation has shown the ability to increase the physical stability of cosmetics [23]. Besides, there may be incompatibility of the nanocarrier with the semi-solid vehicle. Hence, compatibility of ingredients from nanocosmetics can also be performed [19].

Further, other tests can also be applied to nanocosmetics such as extrudability and nanocosmetic particle size determination. The first one determines the ease of formulation be removed from its packaging [19]. The second one detects the presence of nanometric particles in the semi-solid form [24].

3. Biological evaluation of nanocosmetics

Biological evaluation of nanocosmetics comprises efficacy and toxicity tests [12, 17, 19, 35] aiming to demonstrate its performance and safety profile [25]. Firstly, the type of nanocarrier to be prepared must be established. This choice should take into account the properties of each nanocarrier. In such a way, if the nanocosmetic must have a moisturizing effect, lipid nanocarriers may be an option, since they have moisturizing activity [10, 20]. If a sunscreen formulation is intended, lipid nanocarriers can also be prepared [16] due to their light scattering properties.

Regardless of the carrier type, biological assessment can be performed directly on the nanocarriers and also on the nanocosmetics [26]. Sometimes, researchers choose at first to perform tests only with nanocarriers [5, 11, 31, 36] in order to establish their properties and potential applications before adding them into a semi-solid form. In other cases, efficacy tests can be done directly on nanocosmetics [17–20, 23].

Performance assays are selected according to the benefit promoted by nanoencapsulation. Thereby, an appropriate efficacy test for a photounstable substance loaded in the nanocarrier is the photostability test [36]. On the other hand, for a substance with a limited skin distribution profile or with undesirable skin permeation, skin permeation/penetration study may be performed [5, 35, 37]. Furthermore, safety assays aim to demonstrate nanocarrier safety profile when applied to a culture of cells [29, 38].

3.1 Biological evaluation of nanocarriers

Nanocarriers can be submitted to efficacy and safety tests [25] after a proper characterization. Efficacy tests assess its performance. Safety tests such as cell viability sets the nanocarrier cellular toxicity [25, 29]. **Table 4** below summarizes some of the most used methodologies/tests. *In vitro* release and skin permeation/penetration tests are widely used [9, 11, 13, 31] as efficacy tests. Additionally, assays such as cell uptake [5, 38], antioxidant activity [12], moisturizing effect [13], and content of nanoencapsulated substance [24] are also employed as efficacy tests.

In vitro release simulates the substance release, and for this reason, it is widely used. Nanocarriers usually have a better performance than conventional formulations (without nanocarriers), favoring the release of the nanoencapsulated substance [22]. Besides, nanocarriers show an initial burst release followed by a prolonged release [11, 13, 21, 31, 35]. This prolonged action is a desirable effect as it reduces the number of product reapplications. In addition, there are also some differences regarding release profile of each nanocarrier [11, 21] (for example, see [21]). Hence, depending on the objective, it is possible to prepare NLC instead of SNL in order to favor the substance release. Furthermore, variations in nanocarriers components are also alternatives to modulate biological effect. In this sense, quantitative variations in nanocarriers composition affect *in* vitro release and moisturizing effect [13].

Skin penetration/permeation studies are conducted to detect the location of a substance in skin layers (penetration) and/or its presence in the receptor medium (indicative of skin permeation) [5, 36]. Skins from different sources are used (See **Table 3**). In the same way as for *in vitro* release tests, nanocarrier type also impacts the skin distribution profile. NLC promote greater skin retention than nanoemulsions. [6, 11], and nanoemulsions may have a small degree of permeation. However, regardless of nanocarrier type, they promote a greater skin penetration than conventional formulations [8]. In the case of cosmetic products, penetration/retention in the epidermis or dermis is desired, and its permeation (detection in the blood circulation) is unwanted [8]. It is yet noteworthy that nanoencapsulation of vegetable ingredients promotes its better retention in the epidermis and dermis. [5, 9, 35]. As there is a trend of the use of natural ingredients in cosmetics, this result is relevant because it assures an increased efficacy of vegetable ingredients loaded in nanocarriers [5].

Cell uptake is a complementary study to skin permeation allowing to confirm the skin location/deposition of nanoencapsulated substances. In this way,

Biological studies	Methodology	Nanocarrier	References
Penetration/permeation study	Permeation in mice skin	PNC	[38]
		NLC	[9]
	Penetration/permeation in porcine ear skin	NE	[5]
	Penetration study in human skin	NLC	[11]
	Permeation/penetration in albino rats	NLC	[13]
	Permeation in dermis pig ear	SLN, NLC	[8]
	Penetration in stratum corneum/ epidermis from human skin	NLC, NE	[36, 6]
In vitro release	Membrane dialysis	NLC	[31]
		NLC, NE	[11]
		NLC	[13]
		PNC	[22]
		NLC	[21]
	Franz diffusion cell	CSLN	[29]
		NLC	[35]
Cellular uptake	CLSM	PNC	[38]
	CLSM	NE	[5]
	Flow cytometer	CSLN	[29]
Cell viability	Sulforhodhamine assay	PNC	[38]
	MTT	CSLN	[29]
		PNC	[25]
Antioxidant activity	Chemiluminescence	NLC	[31]
		NLC	[12]
	FRAP	NLC, NE	[11]
	DPPH	NLC	[13]
Moisturization effect	Occlusion factor	NLC	[13]
	TEWL and skin hydration	NLC	[10]
Photostability	Content of tocopherol	NLC, NE	[11]
	Content of rose hip oil	PNC	[24]

PNC: polymeric nanocarrier, NLC: nanostructured lipid carriers, SLN: solid lipid nanoparticles, NE: nanoemulsion, CLSM: Confocal laser scanning microscopy, MTT: 3-[4,5-dimethylthiazol-2-yl]-2,5-diphenyl tetrazolium bromide, CSLN: cationic solid lipid nanoparticles, FRAP: ferric reducing antioxidant potential, DPPH: (1,1-diphenylpicryl-hydrazyl), TEWL: transepidermic water loss.

Table 4.
Biological evaluation of nanocarriers.

nanoencapsulation of vegetable ingredients modulates its skin distribution and allows its skin retention as evidenced by confocal microscopy [38]. Besides, a strategy to increase cellular uptake is the use of nanocarriers with positive electrical charge (CSLN) [29].

In contrast, as previously mentioned, cell viability is a safety assay. As nanosystems may cause cell/tissue toxicity, assessing their safety is important. Nanocarriers with low cytotoxicity are considered safety [11, 25, 26]. Regarding other efficacy assays, nanocarriers also have a better performance. Nanoencapsulation of vegetable ingredients increase its antioxidant activity [12, 26]. Nanocarriers also promote skin moisturization [10, 34] where the moisturizing effect depends on the carrier composition [10]. Nanocarriers increase photostability of entrapped substances [11, 24], which affects its antioxidant activity. Since there is a less amount of substance degraded by light, a greater proportion of it will be available to act as an antioxidant.

3.2 Biological evaluation of nanocosmetics

Biological assessment of nanocosmetics comprises the same assays employed to nanocarriers such as penetration/skin permeation, skin moisturization, or antioxidant activity. However, other tests are also used for semi-solid forms such as the sun protection factor (SPF) and ultraviolet A protection factor (UVA-PF) [31]. **Table 5** below exhibits the biological tests applied to nanocosmetics, whether they are efficacy or safety tests.

As described previously, skin penetration/permeation determines substance accumulation in skin layers and its permeation, if any. Penetration and/or permeation will be desirable depending on the type of product. For antioxidants, penetration in the epidermis and dermis is desirable [5]. Besides, ingredients of nanocosmetics also affect skin permeation/penetration, whether they are ingredients of the nanocarrier or the ingredients of semi-solid form [25]. As reported, nanocosmetics have a better skin retention/penetration than conventional cosmetics [26].

SPF and UVA-PF are efficacy tests used to evaluate UVB and UVA protection, respectively [14]. Nanocosmetics have shown superior performances regarding these assays. Cosmetics containing nanocarriers with different quantitative compositions also have different SPF and PF-UVA values [15]. In addition, nanoencapsulation also improves ultraviolet protection (for example, see [16]). Nanocosmetics containing association of vegetable oils have also been used to promote ultraviolet protection [12].

Moisturizing effect is commonly performed *in vitro* [19, 26] or in humans [18, 20] and is usually an important measure for moisturizing and anti-aging cosmetics [20]. Cosmetics containing lipid nanocarriers have moisturizing properties due to their adhesive characteristics, to their composition [18, 20], and to the composition of semi-solid vehicle [17]. Nanocosmetics also preserve antioxidant activity even after irradiation exposure. Once there is depletion of endogenous antioxidants with ultraviolet radiation exposure, these nanocosmetics can be used as anti-aging cosmetics [26]. Another important test is skin toxicity assessment whose goal is to detect any irritation caused by nanocosmetics [19].

Additionally, nanocosmetics have a photostabilizing effect. Nanocosmetics have the ability to reduce the degradation of nanoencapsulated ingredients [23].

Biological assay	References
Skin penetration/permeation	[19, 25, 26]
Sunscreen protection factor (SPF)	[12, 15, 16, 19, 25, 26, 31]
Ultraviolet A protection factor (UVA-PF)	[12, 16, 31]
Skin moisturization	[17–20, 26]
Antioxidant activity	[26]
Toxicity	[19, 25, 26]

Table 5.
Biological assays applied to nanocosmetics.

Multifunctional nanocosmetics have also the ability to increase biological efficacy [14]. There is also a trend toward more complex cosmetics which combines nanocarriers, inorganic nanoparticles, and conventional cosmetic ingredients (without nanotechnology) (for example, see [26]).

4. Conclusion

Nanotechnology-based cosmetics are increasingly common due to their many benefits. Understanding nanocarrier properties is a complex and expensive task that requires professionals with expertise in nanotechnology as well as requiring a high investment in the development of nanotechnology-based products. After deciding the suitable nanocarrier, its composition, and preparation method, characterization must be accomplished. Subsequently, nanocarrier and/or cosmetics containing nanocarriers must be evaluated regarding biological tests to assure an effective and safe product.

Ultimately, there is a trend to use biocompatible ingredients as well as the use of natural ingredients in nanocarriers. An association/co-encapsulation of multifunctional ingredients is also reported. Therefore, multifunctional nanocosmetics represent a great alternative to improve the performance of cosmetics.

Author details

Júlia Scherer Santos
Federal University of Juiz de Fora, Juiz de Fora, Brazil

*Address all correspondence to: julia_scherer_santos@hotmail.com

References

[1] Farris PK, editor. Cosmeceuticals and Cosmetic Practice. Chichester, UK: Wiley; 2013. p. 295. DOI: 10.1002/9781118384824

[2] Cornier J, Keck CM. Nanocosmetics. London: Springer Nature; 2019. p. 262. DOI: 10.1007/978-3-030-16573-4

[3] Mihranyan A, Ferraz N, Strømme M. Current status and future prospects of nanotechnology in cosmetics. Progress in Materials Science. 2012;**57**:875-910. DOI: 10.1016/j.pmatsci.2011.10

[4] Baccarin T, Mitjans M, Ramos D, et al. Photoprotection by Punica granatum seed oil nanoemulsion entrapping polyphenol-rich ethyl acetate fraction against UVB-induced DNA damage in human keratinocyte (HaCaT) cell line. Journal of Photochemistry and Photobiology, B: Biology. 2015;**153**:127-136

[5] Baccarin T, Lemos-Senna E. Potential application of nanoemulsions for skin delivery of pomegranate peel polyphenols. AAPS PharmSciTech. 2017;**18**:3307-3314. DOI: 10.1208/s12249-017-0818-x

[6] Puglia C, Damiani E, Offerta A, et al. Evaluation of nanostructured lipid carriers (NLC) and nanoemulsions as carriers for UV-filters: Characterization, in vitro penetration and photostability studies. European Journal of Pharmaceutical Sciences. 2014;**51**: 211-217. DOI: 10.1016/j.ejps.2013.09.02

[7] Hanno I, Anselmi C, Bouchemal K. Polyamide nanocapsules and nano-emulsions containing Parsol® MCX and Parsol®1789: In vitro release, ex vivo skin penetration and photo-stability studies. Pharmaceutical Research. 2012;**29**:559-573. DOI: 10.1007/s11095-011-0592-5

[8] Mitri K, Shegokar R, Gohla S, et al. Lipid nanocarriers for dermal delivery of lutein: Preparation, characterization, stability and performance. International Journal of Pharmaceutics. 2011;**414**:267-275. DOI: 10.1016/j.ijpharm.2011.05.008

[9] Chen-yu G, Chun-fen Y, Qi -lu L, et al. Development of a Quercetin-loaded nanostructured lipid carrier formulation for topical delivery. International Journal of Pharmaceutics. 2012;**430**:292-298. DOI: 10.1016/j.ijpharm.2012.03.042

[10] Loo CH, Basri M, Ismail R, et al. Effect of compositions in nanostructured lipid carriers (NLC) on skin hydration and occlusion. International Journal of Nanomedicine. 2013;**8**:13-22. DOI: 10.2147/IJN.S35648

[11] Abla MJ, Banga a K. Formulation of tocopherol nanocarriers and in vitro delivery into human skin. International Journal of Cosmetic Science. 2014;**36**:239-246. DOI: 10.1111/ics.12119

[12] Badea G, Lăcătuşu I, Badea N, et al. Use of various vegetable oils in designing photoprotective nanostructured formulations for UV protection and antioxidant activity. Industrial Crops and Products. 2015;**67**:18-24. DOI: 10.1016/j.indcrop.2014.12.049

[13] Ammar HO, Ghorab MM, Mostafa DM, et al. Folic acid loaded lipid nanocarriers with promoted skin antiaging and antioxidant efficacy. Journal of Drug Delivery Science and Technology. 2016;**31**:72-82. DOI: 10.1016/j.jddst.2015.11.007

[14] Lacatusu I, Arsenie LV, Badea G, et al. New cosmetic formulations with broad photoprotective and antioxidative activities designed by amaranth and pumpkin seed oils nanocarriers. Industrial Crops and Products. 2018;**123**:424-433. DOI: 10.1016/j.indcrop.2018.06.083

[15] Niculae G, Lacatusu I, Badea N, et al. Rice bran and raspberry seed oil-based nanocarriers with self-antioxidative properties as safe photoprotective formulations. Photochemical & Photobiological Sciences. 2014;**13**:703-716. DOI: 10.1039/c3pp50290b

[16] Niculae G, Badea N, Meghea A, et al. Coencapsulation of butyl-methoxydibenzoylmethane and octocrylene into lipid nanocarriers: UV performance, photostability and in vitro release. Photochemistry and Photobiology. 2013;**89**:1085-1094. DOI: 10.1111/php.12117

[17] Barua S, Kim H, Hong SC, et al. Moisturizing effect of serine-loaded solid lipid nanoparticles and polysaccharide-rich extract of root Phragmites communis incorporated in hydrogel bases. Archives of Pharmacal Research. 2017;**40**:250-257. DOI: 10.1007/s12272-016-0846-1

[18] Souza C, de Freitas LAP, Maia Campos PMBG. Topical formulation containing beeswax-based nanoparticles improved in vivo skin barrier function. AAPS PharmSciTech. 2017;**18**:2505-2516. DOI: 10.1208/s12249-017-0737-x

[19] Netto MPharm G, Jose J. Development, characterization, and evaluation of sunscreen cream containing solid lipid nanoparticles of silymarin. Journal of Cosmetic Dermatology. 2018;**17**:1073-1083. DOI: 10.1111/jocd.12470

[20] Tichota DM, Silva AC, Sousa Lobo JM, et al. Design, characterization, and clinical evaluation of argan oil nanostructured lipid carriers to improve skin hydration. International Journal of Nanomedicine. 2014;**9**:3855-3864. DOI: 10.2147/IJN.S64008

[21] Bose S, Michniak-Kohn B. Preparation and characterization of lipid based nanosystems for topical delivery of quercetin. European Journal of Pharmaceutical Sciences. 2013;**48**:442-452. DOI: 10.1016/j.ejps.2012.12.005

[22] Ayumi NS, Sahudin S, Hussain Z, et al. Polymeric nanoparticles for topical delivery of alpha and beta arbutin: Preparation and characterization. Drug Delivery and Translational Research. 2019;**9**:482-496. DOI: 10.1007/s13346-018-0508-6

[23] Külkamp-Guerreiro IC, Terroso TF, Assumpção ER, et al. Development and stability of innovative semisolid formulations containing nanoencapsulated lipoic acid for topical use. Journal of Nanoscience and Nanotechnology. 2012;**12**:7723-7732. DOI: 10.1166/jnn.2012.6662

[24] Contri RV, Kulkamp-Guerreiro IC, da Silva SJ, et al. Nanoencapsulation of rose-hip oil prevents oil oxidation and allows obtainment of gel and film topical formulations. AAPS PharmSciTech. 2016;**17**:863-871. DOI: 10.1208/s12249-015-0379-9

[25] Joshi H, Hegde AR, Shetty PK, et al. Sunscreen creams containing naringenin nanoparticles: Formulation development and in vitro and in vivo evaluations. Photodermatology, Photoimmunology & Photomedicine. 2017;**34**:69-81. DOI: 10.1111/phpp.12335

[26] Shetty PK, Venuvanka V, Jagani HV, et al. Development and evaluation of sunscreen creams containing morin-encapsulated nanoparticles for enhanced UV radiation protection and antioxidant activity. International Journal of Nanomedicine. 2015;**10**: 6477-6491. DOI: 10.2147/IJN.S90964

[27] De Oliveira CA, Peres DDA, Graziola F, et al. Cutaneous biocompatible rutin-loaded gelatin-based nanoparticles increase the SPF of the association of UVA and UVB filters. European Journal of Pharmaceutical

Sciences. 2016;**81**:1-9. DOI: 10.1016/j.ejps.2015.09.016

[28] Grumezescu AM. Nanobiomaterials in Galenic Formulations and Cosmetics. Oxford: Elsevier; 2016. p. 432. DOI: 10.1016/C2015-0-00402-6

[29] Jeong YM, Ha JH, Park SN. Cytoprotective effects against UVA and physical properties of luteolin-loaded cationic solid lipid nanoparticle. Journal of Industrial and Engineering Chemistry. 2016;**35**:54-62. DOI: 10.1016/j.jiec.2015.12.014

[30] Mahdi ES, Noor AM, Sakeena MH, et al. Formulation and in vitro release evaluation of newly synthesized palm kernel oil esters-based nanoemulsion delivery system for 30% ethanolic dried extract derived from local Phyllanthus urinaria for skin antiaging. International Journal of Nanomedicine. 2011;**6**: 2499-2512. DOI: 10.2147/IJN.S22337

[31] Lacatusu I, Badea N, Murariu A, et al. The encapsulation effect of uv molecular absorbers into biocompatible lipid nanoparticles. Nanoscale Research Letters. 2011;**6**:73. DOI: 10.1016/j.indcrop.2018.06.083

[32] Zamarioli CM, Martins RM, Carvalho EC, et al. Nanoparticles containing curcuminoids (Curcuma longa): Development of topical delivery formulation. Brazilian Journal of Pharmacognosy. 2015;**25**:53-60. DOI: 10.1016/j.bjp.2014.11.010

[33] Pereira TA, Guerreiro CM, Maruno M, et al. Exotic vegetable oils for cosmetic O/W nanoemulsions: In vivo evaluation. Molecules. 2016;**21**: 1-16. DOI: 10.3390/molecules21030248

[34] De Azevedo Ribeiro RC, Barreto SMAG, Ostrosky EA, et al. Production and characterization of cosmetic nanoemulsions containing Opuntia ficus-indica (L.) mill extract as moisturizing agent. Molecules. 2015;**20**:2492-2509. DOI: 10.3390/molecules20022492

[35] Salunkhe SS, Bhatia NM, Pokharkar VB, et al. Topical delivery of Idebenone using nanostructured lipid carriers: Evaluations of sun-protection and anti-oxidant effects. Journal of Pharmaceutical Investigation. 2013;**43**:287-303. DOI: 10.1007/s40005-013-0079-y

[36] Puglia C, Bonina F, Rizza P, et al. Lipid nanoparticles as carrier for octyl-methoxycinnamate: in vitro percutaneous absorption and photostability studies. Journal of Pharmaceutical Sciences. 2012;**101**: 301-311. DOI: 10.1002/jps

[37] Gilbert E, Roussel L, Serre C, et al. Percutaneous absorption of benzophenone-3 loaded lipid nanoparticles and polymeric nanocapsules: A comparative study. International Journal of Pharmaceutics. 2016;**504**:48-58. DOI: 10.1016/j.ijpharm.2016.03.018

[38] Kwon MC, Choi WY, Seo YC, et al. Enhancement of the skin-protective activities of Centella asiatica L. urban by a nano-encapsulation process. Journal of Biotechnology. 2012;**157**:100-106. DOI: 10.1016/j.jbiotec.2011.08.025

Chapter 7

Chalcones in Dermatology

Jumina, Harizal and Yehezkiel Steven Kurniawan

Abstract

The human skin is pivotal for protecting the body from various stresses and diseases, regulating several physiological aspects, and sensing any signal changes around the environment. To work and function optimally, the skin should be protected and cared regularly by using some treatments. Chalcone, as a privileged structure, exhibits wide and unique bioactivities related to several skin disorders such as in preventing and treating pigmentation disorders (melasma and vitiligo), cutaneous leishmaniasis, rashes (acne vulgaris, seborrheic dermatitis and dandruff, psoriasis and atopic dermatitis), and rosacea. In this chapter, the role of chalcone derivatives in treating several skin disorders as mentioned above is discussed to provide a brief and comprehensive perspective regarding the role of chalcone in dermatology including in vitro, preclinical, and clinical assays.

Keywords: chalcones, dermatology, skin disorder, bioactivity, skin treatment

1. Introduction

In the human body, the skin is the outermost and largest organ with three basic functions, i.e. protecting from various stresses and hazards, regulating some physiological aspects, and sensing any conditional changes in the environment. While conducting its functions, the skin works with other internal systems and forms a rigid network with nervous, immune, and endocrine systems [1]. On the other hand, the skin also interacts with the environment around [2] such as microorganisms living on the skin surface [3] to maintain the function of the human body. Any imbalances between these factors often lead to various multifactorial skin disorders [4]. Based on their main cause, skin disorders can be classified into six groups including tumor and cancer, trauma, pigmentation disorders, microbial (viral, bacterial, fungal, and parasitic) infections, rashes, and miscellaneous conditions [5]. Therefore, treatment of skin disorders requires systematic attention in the medical field.

The treatment of skin disorders is usually conducted in topical, systematic, or combinatorial modes using bioactive compounds to relieve any formed defects in nervous, immune, and endocrine systems. Various bioactive compounds have been evaluated through a long clinical assay [6]. In the development process of these compounds, many parameters should be considered such as economical aspect, bioavailability, stability, toxicity, and metabolism of drug molecules in both topical and systematic therapies [7].

Chalcone is a class of organic compounds with 1,3-diaryl-2-propen-1-one (**Figure 1**) as the backbone structure that is obtained either through an isolation and purification process from natural samples [8], semisynthesis process from existing natural products [9], or total synthesis process [10]. Chalcone is considered

IntechOpen

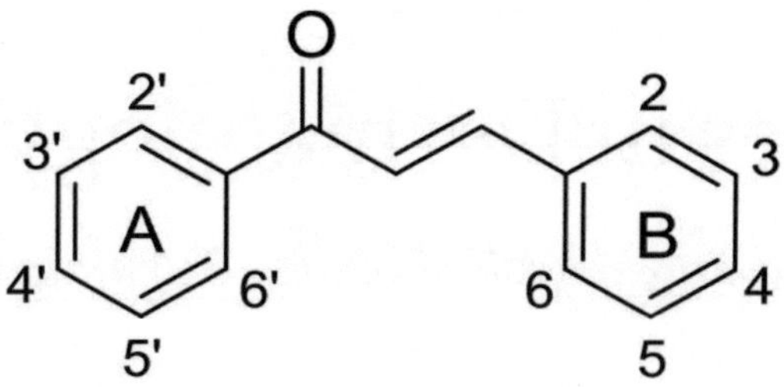

Figure 1.
Backbone structure of chalcones.

as privileged scaffold since it exhibits broad biological activities [10]. Many pieces of study have used chalcone as a lead compound to find more potent drugs for diseases due to the ease and convenience in synthesizing and functionalizing chalcone structure. In treating skin disorders, the efficacy of chalcone derivatives is related to several biological activities including antioxidant [11], anti-inflammation [12], immunomodulation [13], anti-angiogenesis [14], antimicrobial [15], and enzyme modulator [16]. The usage of chalcone derivatives in dermatology has been developed over the past several years to obtain the most active ingredient in either for maintaining skin health or treating skin disorders. Licochalcone A, isoliquiritigenin, and xanthohumol are examples of well-known chalcone derivatives with low toxicity and side effects on treating skin disorders and disease. In this chapter, these applications are described and discussed to provide a broad and comprehensive perspective regarding the role of chalcones in the dermatology field.

2. Dangerous impacts of ultraviolet irradiation

The ultraviolet (UV) light is the electromagnetic radiation with a wavelength range from 200 to 400 nm. In general, based on the wavelength range, UV radiation is divided into four regions including UVC, UVB, UVA2, and UVA1 at 200–280, 280–320, 320–340, and 340–400 nm, respectively. This hazardous radiation causes various acute and chronic negative effects on the human skin. Clinical manifestations that occur usually depend on the wavelength and intensity of UV radiation, part of the body exposed, and type of skin (based on Fitzpatrick's classification of skin). Acute effects of UV radiations, in general, involve various forms of inflammations such as erythema, local immunosuppression, phototanning, and epidermis thickening [17]. Mechanistically, acute effects are initiated by suitable interactions between several chromophores (either on the epidermis or dermis layers) and UV irradiation [18]. These interactions trigger the structural changes that modulate various biochemical and immunological processes [19] such as releasing several pro-inflammatory cytokines, damaging various cell biomolecules [20], generating various reactive oxygen species (ROS) [21], and producing several inflammatory mediators such as prostaglandins, histamine, and leukotrienes [22, 23]. Meanwhile, chronic effects usually lead to photoadaptation or photoprotection effects through the formation of photoaging, immunosuppression, and photocarcinogenesis [17].

2.1 Chalcones as sunscreen active ingredients

Chalcone has been found as one of the bioactive compounds that is able to reduce negative effects from UV radiation such as hesperidin methyl chalcone, licochalcone A, etc. Chalcones as antioxidant and anti-inflammatory agents are used in topical or systemic methods, while the sunscreen agent is used in topical applications. In general, the sunscreen activity of chalcones is generated by high UV absorbance on the UVB-UVA region [24], which corresponds to $n \rightarrow \pi^*$ and $\pi \rightarrow \pi^*$ electronic transition from HOMO to LUMO energy levels in chalcones' conjugated electronic system. The absorbance region of chalcones can be shifted by introducing electron-donor substituents (bathochromic shift) or electron-withdrawing substituents (hypsochromic shift) in both of the chalcone aromatic rings. However, the effects of these groups are more significant if present in A ring compared to B ring [25].

The main problem in using chalcone derivatives as sunscreen active compound is relatively poor photostability and transformations into various by-products [26–28]. Several reports have investigated the utilization of chalcone derivatives as sunscreen active compounds and also prepared sunscreen formulations with high compatibility on the skin. However, the photostability of chalcone derivatives used has not been evaluated yet [25, 26].

2.2 Chalcones as photoprotective agents

Several chalcones exhibit strong protection on UV-induced deleterious effects such as *trans*-chalcone, butein, monspermoside, licochalcone A, phloretin, and hesperidin derivatives. Unsubstituted *trans*-chalcones showed a potential activity in reducing inflammation effect and oxidative stress in mice [27, 28]. A formulation containing 1% unsubstituted *trans*-chalcone has been applied to protect the skin from UVB radiation by inhibiting inflammation through reducing tumor necrosis factor alpha (TNF-α) levels and improving antioxidant and detoxification systems through enhancing heme oxygenase 1 (HO-1) and nuclear factor erythroid 2-related factor 2 (Nrf2) messenger ribonucleic acid (mRNA) expressions [28]. Systemic administration of *trans*-chalcone could inhibit UVB-induced skin inflammation and prevented oxidative stress by targeting nicotinamide adenine dinucleotide phosphate H (NADPH) oxidase and cytokine production [27]. Butein and monspermoside compounds were also used as photostabilizer for UVA-absorbing compounds such as dibenzoylmethane [29].

Licochalcone A, isolated from *Glycyrrhiza inflata*, is the most well-studied chalcone derivative related to its activity as UV photoprotector. Either in vitro or in vivo studies showed that licochalcone A had a strong protective effect against UVB-induced oxidative stress and inflammation. Licochalcone A attenuated UVB-induced inflammation by inhibiting prostaglandin E2 (PGE2), cyclooxygenase (COX-2), lipoxygenase, nuclear factor kappa-light-chain-enhancer of activated B cells (NF-κB), and Nrf2 [30–32], while in vivo assay showed that topical formulation containing licochalcone A caused a significant reduction in UV-induced erythema, irritation, and oxidative process in the skin [31–33].

Similar to their flavanone analog, hesperidin [34], hesperidin methyl chalcone, and hydrolyzed methylhesperidin compounds as semisynthetic products also exhibit high UV protective effect. Topical and systematic administration of hesperidin methyl chalcone in hairless mice inhibited UVB-induced oxidative stress by reducing free radicals and ROS, enhanced endogenous antioxidant systems, and inhibited inflammation by reducing the production of cytokines [35]. The hydrolysis product of methylhesperidin, 4′,6′-dihydroxy-3,4,2′-trimethoxychalcone and 4′-hydroxy-3,4,2′,6′-tetramethoxychalcone compounds, also induced cytoprotective gene expression and reduced oxidative stress by promoting Nrf2 nuclear translocation and antioxidant response element (ARE) luciferase activity in UVB-irradiated keratinocytes [36].

Phloretin is a natural dihydrochalcone that exhibited a strong inhibition of several matrix metalloproteinases. The isolated 3-hydroxyphloretin and phloretin from *Malus doumeri* var. *formosana* showed high inhibition of MMP-1 production in fibroblast cells [37]. A combination of phloretin, ferulic acid, and vitamin C as antioxidants had a high protective effect on UV damage in the human skin by preventing erythema formation and inhibiting MMP-9 and thymine dimerization reaction. In this case, phloretin acted as an antioxidant and synergized with other antioxidants by stabilizing and enhancing the bioavailability of ferulic acid and vitamin C in the skin [38].

3. Pigmentation disorders

Melanogenesis is a complex process of production and distribution of melanin by melanocytes. In this process, melanins (in form of eumelanin, pheomelanin, or trichochrome) are synthesized and stored in melanosomes (an organelle in melanocytes) and then transported to nearby keratinocytes to act as photoprotector in the skin and lead to chronic pigmentations [39]. The synthesis of melanin involves several reaction steps and is catalyzed by phenylalanine hydroxylase (PAH), tyrosinase (TYR), tyrosine hydroxylase isoform I (THI), tyrosinase-related protein 1 (TRYP1), and tyrosinase-related protein 2 (TRYP2) [40] enzymes. The rate-determining step of this process is hydroxylation of L-tyrosine to L-dopaquinone catalyzed by tyrosinase enzyme [41]. The final products of this process include black-brown eumelanin, yellow-reddish brown pheomelanin, and trichochrome [42].

There are many internal (endocrine, immune, inflammatory, and central nervous systems) and external (ultraviolet radiation and drugs) factors that affect the melanogenesis process [41]. Any disruptions from these factors will cause different types of pigmentation disorders including various skin conditions with strange melanocyte density, melanin concentration, or both that change pigmentations in the skin [43]. In general, pigmentation disorders can be divided into two main groups including hyperpigmentation and hypopigmentation that related to the amount of changes of normal melanin pigmentation, respectively. Both of these groups include several diseases with different clinical manifestations, but the most common forms of these are melasma (hyperpigmentations) and vitiligo (hypopigmentations) [42, 43].

3.1 Melasma

Melasma is a multifactorial skin disease indicated by the presence of symmetrical hyperpigmented area in a certain part of the face including centrofacial, malar, and mandibular parts. This disease is affected by UV, visible, and infrared exposure, by inducing reactive oxygen species and promoting melanogenesis [44], hormonal conditions [45], and genetics [46]. Chalcone derivatives are extensively used in medical therapy as photoprotective described in Sections 2.1 and 2.2, and hypopigmenting agents. Various studies showed that either natural or synthetic chalcones exhibited strong activity in inhibiting cellular tyrosinase and reducing cellular melanin formation [47]. However, the action mode of chalcone derivatives in this study is unknown since most of these studies used mushroom tyrosinase that is different from human tyrosinase [48]. Other studies conducted by Kim and coworkers showed similar results with additional parameters. In these studies, chalcones containing cyclohexylmethoxy group not only attenuated cellular melanin production and tyrosinase activity, but also reduced the expressions of several melanogenesis-related genes (transcriptional activity of tyrosinase and microphthalmia-associated transcription factor/MITF) and proteins (TRP1, TRP2, and MITF) [49, 50].

Table 1 shows several in vitro assays that use Murine B16 melanoma cell lines (B16F10), melanoma cells, and human melanocyte (G361) cells in determining chalcone derivatives' efficacy for hypopigmenting agent. In vivo studies also showed that licochalcone A and isoliquiritigenin-containing licorice extract cream can improve melasma and increase skin brightness [51]. Loading the licorice extract into solid lipid nanoparticles has been formulated and applied in a clinical trial for melisma [52].

3.2 Vitiligo

Vitiligo is an acute skin disease characterized by unilaterally distributed depigmented areas in the skin due to the systematic degradation of melanocytes [56]. The mechanism triggering this disease has not completely elucidated yet; however, there are several theories proposed to explain it including genetics (mutations of certain genes cause autoimmunity) [57], autoimmunity (instable melanocytes induce immune system activation), oxidative stress (endogenous and exogenous stress-induced reactive oxygen species that cause internal damage of melanocytes)

Chalcone derivatives	Source	Types of cells	Effects (ref.)*
Isoliquiritigenin	*Glycyrrhiza glabra* L.	G361	↓Cellular melanin formation (IC_{50} 4.73 μg/mL) [53]
2,4,3′,4′-Tetrahydroxychalcone	Synthetic	G361	(M = 5 μM) ↓cellular melanin content (62.5%) [54]
2,4,2′,4′-Tetrahydroxychalcone	Synthetic	G361	(M = 5 μM) ↓cellular melanin content (55.0%) [54]
3-Hydroxyphloretin	*Malus doumeri* var. *formosana*	HEMn**	(M = 100 μM) ↓cellular tyrosinase (80.5%) and cellular melanin content (18.3%) [55]
2′-Cyclohexylmethoxy-6′-hydroxy-4-hydroxymethylchalcone	Synthetic	B16F10	↓Cellular melanin production (IC_{50} = 6.2 μM) and cellular tyrosinase activity (IC_{50} = 6.8 μM); ↓TRP1, TRP2, and MITF expression; ↓transcriptional activity of tyrosinase (IC_{50} = 6.0 μM) and MIRF (IC_{50} = 5.8 μM) [49]
4-Acetamido-2′-cyclohexylmethoxychalcone	Synthetic	B16F10	↓Cellular melanin production (IC_{50} = 0.54 μM) and cellular tyrosinase activity; ↓MITF, TRP1 and TRP2 expression; ↓phosphorylation of ERK1/ERK2 and CREB; ↓transcriptional activity of MITF and CRE [50]

**IC50, half maximal inhibitory concentration; ERK1/ERK2, extracellularly regulated kinase 1 and 2; CREB, cyclic adenosine monophosphate (cAMP) response element-binding protein.*
***HEMn: Primary skin melanocyte cells from neonatal foreskin.*

Table 1.
Chalcone derivatives activities in inhibiting cellular melanogenesis.

Chalcone derivatives	Source	Types of cells	Effects (ref.)
Chalcone-containing Kaliziri extract	*Vernonia anthelmintica* (L.) Willd.	B16F10	↑Tyrosinase, TRP-1, TRP-2, and MITF expression [64]
4′-(3-(3,4-Difluorophenyl)isoxazol-5-yl)methoxychalcone	Synthetic	B16F10	(M = 50 μM) ↑cellular melanin content (463%) [62]
4-(3-(2,3-Dihydrobenzo[b][1,4]dioxin-6-yl)isoxazol-5-yl) methoxychalcone	Synthetic	B16F10	(M = 50 μM) ↑cellular melanin content (438%) [62]
4′-Methoxy-4-dimethylaminochalcone	Synthetic	B16F10	(M = 40 μM) ↑cellular melanin content (75%); ↑cellular tyrosinase activity (30%) [65]

Table 2.
Chalcone derivatives' activities in activating cellular melanogenesis.

[58], melanocyte growth and defective melanocyte adhesion (repeated pressure and friction cause detachment of melanocytes to surrounding structures) [59], viral infections (certain viruses cause vitiligo) [60], and neural mechanism (neuropeptides elevate in vitiligo lesions) [61].

Improvement of skin appearance in vitiligo can be approached either by pigmentation or depigmentation. Chalcone derivatives are able to act as hypopigmenting (as described in Section 4.1) and hyperpigmenting agents. Several chalcone derivatives acted as hyperpigmenting agent by activating tyrosinase enzyme and elevating melanin production (**Table 2**). The presence of certain electron-withdrawing substituents such as halogen and trifluoromethyl substituents on the chalcone structure exhibited a significant effect in activating tyrosinase [62]. Chalcones also showed strong absorption in UV regions with low toxicity and have been formulated as broad-spectrum sunscreen protecting the affected skin from UV radiations and photofilter used in narrow-band UVB (NB-UVB) phototherapy [63].

4. Cutaneous leishmaniasis

Leishmaniasis is one of the vector-borne diseases generated by *Leishmania* spp. protozoans and transmitted to mammals through infected female sandflies (*Phlebotomus* and *Lutzomyia*) [66]. In general, certain *Leishmania* species can cause different clinical features with different degree of severity as the result of the interplay between *Leishmani*a species characteristics, biological vector, and the responses of host immune system [67]. In this section, the explanation is focused on the cutaneous leishmaniasis as the most well-known form [68].

Cutaneous leishmaniasis is identified by the presence of skin lesions (ulcers) in the biting spot of *Leishmania* spp.-infected sandfly. Almost all pathogenic *Leishmania* spp. could cause cutaneous leishmaniasis (18 from 20 species) [69] with 24 species of *Phlebotomus* spp. and 40 species of *Lutzomyia* spp. acting as a vector or potential vector [70]. At first, after biting by the infected female sandfly, the spot will form small erythema developing into a papule and then a nodule and ulcerate to become skin lesion. In the human body, promastigotes of *Leishmania* parasites injected by sandfly will be phagocytosed by macrophages [71]. Promastigotes manipulate macrophages to develop and multiply promastigote into amastigotes that infect another sandfly by biting infected human [72]. These processes cause

Species	Compounds	Source	Activity (IC_{50}) and mechanism*
L. major	Licochalcone A**	*Glycyrrhiza uralensis*	↓Intracellular amastigoten (0.5 μg/mL) by damaging amastigote mitochondria and disturbing its function [73] and inhibiting fumarate reductase [74]
L. amazonensis	4-Fluoro-2′-hydroxy-4′,6′-dimethoxychalcone**	Synthetic	↓Promastigote (0.8 μM), ↓intracellular amastigoten (4.3 μM). Mechanism is not related to the inhibition of fumarate reductase [78]
	3′,4′,5′-Trimethoxy-3-nitrochalcone in nanoemulsions	Synthetic	↓Intracellular amastigoten (0.32 μM) [79]
	trans-chalcone	Synthetic	↓Axenic amastigoten (10.3 μM), ↓promastigote (10.3 μM) by ↑ROS production and ↓mitochondrial integrity, phosphatidylserine exposure, and damaging the membrane. Immunomodulator by ↓ TNF-α, TGF-β, IL-10, ROS and NO, ↑ Nrf2, heme oxygenase, and ferritin [80]
	2′,4′-Dihydroxychalcone**	Synthetic	↓Promastigotes (0.4 μM) by inhibiting glycerol-3-phosphate dehydrogenase [77]
	Lonchocarpine/4-hydroxylonchocarpine (3:1)**	*Dorstenia mannii*	↓Intracellular amastigotes (6.64 μg/mL). Activity comes from synergistic effect from two compounds [81]
	2′-Hydroxy-4′,6′-dimethoxy-3-nitrochalcone	Synthetic	↓Intracellular amastigotes (7.2 μg/mL) [82]
L. braziliensis	*trans*-chalcone	Synthetic	↓Promastigotes (1.58 μM) [83]
	4-Methoxy-3-(*N*-phenylsulfamoyl)chalcone	Synthetic	↓Promastigotes (3.50 μM) [84]
	3-Chloro-2′,4′,6′-trimethoxychalcone	Synthetic	↓Promastigotes (2.70 μM) [85]
L. infantum	*(E)*-3-(5-nitrofuran-2-yl)-1-(4-(piperidin-1-yl)phenyl) prop-2-en-1-one**	Synthetic	↓Amastigotes (6.2 μM) by inhibiting cysteine proteases such as procathepsin L [76]
	(E)-3-(3-(3,5-di-tert-butyl-4-hydroxyphenyl) acryloyl)-4-hydroxyquinolin-2(1H)-one**	Synthetic	↓Intracellular amastigotes (1.3 μM) [86]
L. panamensis	*(E)*-1-(4-chlorophenyl)-3-(3-((7-chloroquinolin-4-yl)amino)phenyl) prop-2-en-1-one**	Synthetic	↓Amastigotes (0.79 μg/mL) [87]

**TGF-β, transforming growth factor beta; IL-10, interleukin-10.*
***Been reviewed by de Mello and coworkers [88].*

Table 3.
Several chalcone derivatives with high activity and selectivity against Leishmania *spp.*

the complex inflammatory responses that mediate and determine the appearance of clinical features and the severity degree.

Various investigations have been conducted and showed that several chalcone (natural and synthetic) derivatives have high activity (IC_{50} < 10 μg/mL or IC_{50} < 5 μM) and selectivity against *Leishmania* parasites in vitro and in vivo. However, the mode of action and molecular target of these chalcone derivatives have not been well elucidated yet. Licochalcone A showed strong activity against *L. major* (IC_{50} = 2.4 μg/mL) by damaging ultrastructure of promastigote and amastigote mitochondria of the parasite selectively and disturbing its function as respiration organelle [73]. Several studies also showed that fumarate reductase [74], nucleoside hydrolase, dihydroorotate dehydrogenase, oligopeptidase B and methionyl-*t*RNA synthetase in *L. major* [75], cysteine proteases in *L. infantum* [76], and glycerol 3-phosphate dehydrogenase (G3PDH) of *L. mexicana* [75, 77] can be targeted to kill the parasite. Chalcone derivatives strongly inhibiting these enzymes showed high activity in killing the parasite (**Table 3**). When the sunscreen activity of chalcones is generated by their conjugated electronic system, the antimicrobial activity of chalcones is caused by the presence of halogen, methoxy, hydroxy, and other functional groups.

5. Rashes

5.1 Acne vulgaris

Acne vulgaris, one of multifactorial acute inflammatory diseases, affects the pilosebaceous unit (hair follicles) in the skin. This disease is commonly found in the area with high-density pilosebaceous units (face, neck, upper chest, shoulders, and back) and characterized by the presence of seborrhea (excessive grease production), noninflammatory lesions (open comedones/blackhead and closed comedones/whitehead), inflammatory lesions (papules and pustules), and various degrees of scarring [89].

Pathogenesis of acne involves the interplay between four main factors including hyperseborrhoea (excessive sebum production) mediated by certain androgens and alterations in sebum fatty acid composition, hyperkeratinization within the follicle that lead to the formation keratin plug (microcomedone), pilosebaceous unit colonization by *Cutibacterium acnes* (aerotolerant anaerobic bacterium) colony, and the release of inflammatory mediators in response to the presence of *Cutibacterium acnes* [90]. It was shown that *Staphylococcus epidermidis* (facultative anaerobic bacterium) also had a beneficial role by limiting the colonialization of *Cutibacterium acnes* and inflammation [91].

Several chalcone derivatives exhibited anti-acne activity with mainly as antibacterial and anti-inflammatory agents. Licochalcone A as a potent anti-inflammatory agent [31] has been studied and combined with other active compounds in several anti-acne formulations. Skin care formulation containing licochalcone A, L-carnitine, and 1,2-decanediol can reduce pustule lesions, popular lesions, total lesions, and sebum levels [92]. A combination of 0.1% adapalene gel and moisturizer containing licochalcone A, L-carnitine, and 1,2-decanediol also showed synergistic effect in reducing inflammatory lesions without interfering the efficacy of each active ingredient [93]. Another combination between 0.1% adapalene gel and moisturizer containing licochalcone A, glycolic acid, salicylic acid, and gluconolactone even showed better results than monotherapy using only adapalene [94]. Licochalcone A, in another report, had antibacterial activity against *Cutibacterium acnes* and inhibited *Cutibacterium acnes*-mediated NLRP3 (nucleotide-binding domain, leucine-rich-containing family, pyrin domain-containing-3) inflammasome activation in the skin [95].

Antibacterial effects against *Cutibacterium acnes* have been shown by phloretin, xanthohumol, panduratin A, isopanduratin A, and some chalcone derivatives. Phloretin, a natural dihydrochalcone derivative, showed both antibacterial effects by inhibiting the growth of *Cutibacterium acnes* (MIC = 0.5 mg mL^{-1}) by blocking the activity of KAS III [beta-ketoacyl-(acyl-carrier-protein) synthase III] in *Cutibacterium acnes* and anti-inflammatory effect by attenuating COX-2 and PGE_2 expressions and inhibiting c-Jun N-terminal kinases (JNK) via toll-like receptor 2 (TLR2)-mediated inflammatory signaling in *Cutibacterium acnes*-induced inflammation of the skin [96, 97]. Four natural chalcone derivatives, xanthohumol isolated from *Humulus lupulus* L. (minimum inhibitory concentration/MIC = 0.003 mg mL^{-1}) [98], panduratin A and isopanduratin A isolated from *Kaempferia pandurata* Roxb. (MICs = 0.002 and 0.004 mg mL^{-1}) [99], and 2′,6′-dihydroxy-3′-methyl-4′-methoxydihydrochalcone isolated from *Eucalyptus maculata* (MIC = 0.002 mg mL^{-1}), also exhibited antibacterial activity against *Cutibacterium acnes* [100]. Different mode of action was shown by two synthetic chalcone derivatives, 2,2′-dihydroxychalcone and 2′-hydroxy-2,3,5′-trimethoxychalcone, to reduce sebum secretion and pore size on the skin [101].

5.2 Seborrheic dermatitis and dandruff

Seborrheic dermatitis is a skin illness mainly influencing skin area with high amounts of sebaceous glands (such as the face, scalp, central chest, and anogenital) and characterized with the presence of erythematous patches with superficial scaling. The mildest form of seborrheic dermatitis that only affects the scalp area with no overt inflammation is usually considered as dandruff or pityriasis capitis. In general, seborrheic dermatitis affects adolescents during puberty to adulthood; however, another form of this disease is considered as infantile seborrheic dermatitis affecting babies and young children [102].

The main mechanism triggering this disease remains unknown; however, it is probably caused by the imbalance of three factors: sebum oversecretion by the sebaceous gland that is affected by certain hormonal [103] and environmental [104] conditions, sebum metabolism by *Malassezia* spp. that accumulate linoleic acid, and individual susceptibility to the penetration of the unsaturated fatty acids that mediate pain and itch [105, 106]. The lipid layer of *Malassezia* spp. also stimulated keratinocytes to produce anti-inflammatory (IL-4, IL-10) and pro-inflammatory (IL-1α, IL-6, IL-8, IL-12, TNF-α) cytokines [107, 108].

Licochalcone A is an active anti-inflammatory drug used in several formulations in treating seborrheic dermatitis. Tonic containing urea, lactate, polidocanol, and licochalcone A-containing *Glycyrrhiza inflate* root extract improved dry, itchy, and inflamed scalp condition which is often associated with seborrheic dermatitis. A combination of rinse-off shampoo (containing piroctone olamine and climbazole) and leave-on tonic (containing piroctone olamine and licochalcone A) significantly reduced dandruff and relieved its micro-inflammation. Treatment using this combination also significantly decreased some cytokines including IL-1ra/IL-1α and IL-8, compared to placebo shampoo [109]. In another report, a moisturizer containing licochalcone 0.025% also had a comparable cure rate with hydrocortisone 1% moisturizer in treating infantile seborrheic dermatitis [110].

5.3 Psoriasis

Psoriasis is a multifactorial skin disease characterized by the presence of red, dry, itchy, and scaly plaques due to chronic hyperproliferative keratinocyte cells and inflammatory cell infiltration [111]. Pathogenesis of this disease includes several steps triggered by environmental [111] and genetic [112] factors: (1) secretion of TNF-α by keratinocyte cells triggered by certain environmental and genetic factors, (2) TNF-α-induced dendritic cell activation and production of IL-23, (3) IL-23-induced T-helper 17 (Th17) cell differentiation, and secretion of IL-17A that cause hyperproliferation of keratinocyte cells [113]. Moreover, IL-17A also synergizes with IL-36 secreted by keratinocytes to amplify and support chronic characteristic of this disease [114]. Several molecular targets have been identified with drug candidates being tested in clinical stages such as cytokines (TNF-α, IL-17A, IL-17E, IL-17F, and IL-23 p19 subunit), phosphodiesterase 4, A_3 adenosine receptor, and Janus kinases (JAK1-JAK3) signaling pathways [115].

Recent studies report that chalcones are efficient in improving some psoriasis models. Intraperitoneal injection of isoliquiritigenin compound remedied mouse with psoriasis-like skin predisposition and decreased several pro-inflammatory cytokines (IL-6 and IL-8) as well by inhibition of NF-κB on in vitro and in vivo assays [116]. The isolated licochalcone B and echinatin compounds from *Glycyrrhiza glabra* also exhibited higher activity than isoliquiritigenin compound in relieving psoriasis-like inflammation induced by several pro-inflammatory agents such as 12-*O*-tetradecanoylphorbal-13-acetate (TPA), adenosine diphosphate (ADP), and UVB radiation [117]. In other targets, several α-bromo- and α-tetrafluoromethyl-2′,3,4,4′-tetramethoxychalcones showed anti-inflammatory and immunomodulatory activities through inhibition on IL-3 and interferon-α (IFNα)-induced JAK2/STAT5 (signal transducer and activator of transcription 5) [118] and STAT1 and STAT2 [119] signaling pathways. Furthermore, both compounds were also able to inhibit NF-κB and to activate Nrf2 transcriptional activity [120].

5.4 Atopic dermatitis

Atopic dermatitis (atopic eczema) is a complex inflammatory skin disease characterized by the presence of skin dryness with intense itching sensation, younger/early-onset, and atopy [121]. There are three main factors that contribute to pathogenesis of atopic dermatitis, i.e., (1) genetic and immune factors on the defect of physical and chemical barrier in the skin [122], (2) hyperactive immune cells due to a biological response (mutation on caspase recruitment domain family member 11/CARD11 gene) because of the incoming pollutants and pathogens [123], and (3) environmental factors that contribute in triggering and enhancing severities of atopic dermatitis process [124].

Chalcone-containing formulations have been prepared and evaluated in treating atopic dermatitis. These chalcone derivatives were combined with other active ingredients, and it was found that they acted as anti-inflammatory and antioxidative agents. It was also reported that licochalcone A-containing moisturizer showed slightly lower effect than hydrocortisone lotion in treating mild-to-moderate atopic dermatitis in children [125]. Another study also found that oil-in-water formulation containing *Glycyrrhiza inflata* root extract (licochalcone A), decanediol, menthoxypropanediol, and ω-6-fatty acids had quite similar efficacy compared to hydrocortisone lotion (standard medication) in a 1-week clinical study with significant improvement in skin conditions, which is remarkable [126]. Using 0.02% triamcinolone acetonide cream as a reference, a combination of 4-*t*-butylcyclohexanol and licochalcone A compounds in moisturizer formulation showed a slower improvement rate, but better result in relieving erythema and increasing skin hydration [127]. Meanwhile, water-in-oil emollient containing licochalcone A, ω-6-fatty acids, ceramide 3, and glycerol formulation could serve a preventing effect on reducing atopic dermatitis flares even after stopping the treatment [128].

Other compounds such as isoliquiritigenin and 3′-isopentenyl-2,2′,4,4′-tetrahydroxy-6′-methoxychalcone (ITC) showed similar effects on atopic dermatitis skin. Evaluation of isoliquiritigenin in relieving atopic dermatitis-like lesion in mice caused a significant improvement in skin condition and attenuated several pro-inflammatory parameters such as Immunoglobulin E (IgE), Th2 cytokine upregulation, TNF-α, IL-6, and IL-4 [129]. ITC isolated from *Sophora flavescens* Aiton also ameliorated atopic dermatitis-like model in mice by inducing HO-1 expression which leads to suppression of Th2 chemokine expressions [130].

6. Rosacea

Rosacea is a multifactorial chronic inflammatory skin disorder characterized by the presence of persistent or periodical redness and several kinds of changing in phymatous in the central facial skin (cheeks, chin, nose, and central part of forehead) [131]. Based on the appearance of certain major phenotypes, rosacea is divided into erythematotelangiectatic rosacea, papulopustular rosacea, glandular/hyperplastic rosacea, ocular rosacea, and other special forms such as rosacea conglobata, rosacea fulminans, gram-negative rosacea, steroid-induced rosacea, granulomatous (lupoid) rosacea, Morbihan's disease, and rosacea in children [132]. Pathogenesis of rosacea has not been fully comprehended yet, but there are several factors that involve including (1) genetics, Haber's syndrome with rosacea as one of clinical feature can be inherited; (2) environment, certain environmental factors can trigger rosacea including extreme air temperature, sudden temperature changes, food (caffeine, alcohol, hot and spicy food), sunlight (UV and IR radiation), etc.; (3) overproduction of antimicrobial peptides (AMPs) by congenital immune system such as LL-37; (4) ROS-induced inflammation produced by adaptive immune cells; (5) overexpression of toll-like receptors (TLRs); (6) *Demodex folliculorum*-induced inflammations; and (7) neuroinflammation and vascular hyperactivity [132, 133].

Hesperidin methyl chalcone, licochalcone A, and tetracarboxymethyl naringenin chalcone exhibit significant activity in improving the skin with rosacea. Clinical studies showed that hesperidin methyl chalcone-containing topical formulations could improve infected skin condition by decreasing the proportion of dilated vessels, total vessel area, and IL8 productions. These formulations showed a complementary effect between each active ingredient in relieving inflammation and reducing the redness [134]. Another study reported that hesperidin methyl chalcone had anti-inflammatory and anti-analgesic activities which are two targets in treating rosacea by inhibiting transient receptor potential vanilloid type 1 (TRPV1), oxidative stress, TNF-α, interleukin (IL) production (IL-1β, IL-6, and IL-10), and NF-κB activity [135].

Licochalcone A is a natural product from *Glycyrrhiza inflate* that shows potent activity in treating especially in mild and moderate symptoms. In vivo studies showed that skin care formulations containing licochalcone A provided various activities such as UVA/UVB protecting, moisturizing, and redness concealing abilities that improved skin appearance with rosacea. These formulations exhibited high compatibility with sensitive skin and could be combined with other treatments such as metronidazole treatment [136]. Licochalcone A-containing moisturizer formulation also increased skin hydration and reduced transepidermal water loss [137]. The most recent chalcone derivative used as anti-rosacea is a stabilized form of naringenin chalcone, i.e., tetracarboxymethyl naringenin chalcone (TNC). TNC was obtained from naringenin chalcone by total etherification reaction of methylchloroacetate. In vitro study showed that TNC significantly reduces LL-37, calcitriol, and several LL-37-induced inflammatory mediators in keratinocytes. Clinical test also showed that formulation containing TNC as a single active ingredient reduced the redness of the skin with rosacea compared to untreated skin areas [138].

7. Conclusions

Most of the skin diseases involve any defects in the skin due to its interaction with UV irradiation, pollutant, and/or other internal and external factors. Because of that, several skin diseases such as pigmentation disorders (melasma and vitiligo), cutaneous leishmaniasis, rashes (acne vulgaris, seborrheic dermatitis and dandruff, psoriasis, and atopic dermatitis), and rosacea have been reported and investigated for better medical treatment. Chalcones, a group of privileged molecules with 1,3-diaryl-2-propen-1-one backbone, exhibit high efficacy in treating those skin diseases. These efficacies are strongly related to the good activity of chalcones as antioxidant, anti-inflammation and immunomodulation, anti-angiogenesis, antimicrobial agents, as well as their ability in modulating various enzymes. Even though chalcones are really potential to be used as a photoprotective agent, the utilization of designed synthetic chalcones is still limited due to their low photostability and medium to high toxicity and also because of unknown protein targets of those skin diseases as for today reports. Development in these fields is still required to obtain more potent drugs with excellent biocompatibility and other desired properties.

Acknowledgements

Financial support from the Ministry of Research, Technology and Higher Education of the Republic of Indonesia who has funded the research in chalcones as sunscreen and NB-UVB phototherapy is acknowledged.

Conflict of interest

The authors state that there is no conflict of interest.

Author details

Jumina[1*], Harizal[2] and Yehezkiel Steven Kurniawan[3]

1 Department of Chemistry, Faculty of Mathematics and Natural Science, Universitas Gadjah Mada, Yogyakarta, Indonesia

2 Department of Pharmacy, Faculty of Health Sciences, Universitas Esa Unggul, Jakarta, Indonesia

3 Ma Chung Research Center for Photosynthetic Pigments, Universitas Ma Chung, Malang, Indonesia

*Address all correspondence to: jumina@ugm.ac.id

References

[1] O'Sullivan RL, Lipper G, Lerner EA. The neuro-immuno-cutaneous-endocrine network: Relationship of mind and skin. Archives of Dermatology. 1998;**134**(11):1431-1435

[2] Singh B, Maibach H. Climate and skin function: An overview. Skin Research and Technology. 2013;**19**(3):207-212

[3] Byrd AL, Belkaid Y, Segre JA. The human skin microbiome. Nature Reviews. Microbiology. 2018;**16**(3):143-155

[4] Dainichi T, Hanakawa S, Kabashima K. Classification of inflammatory skin diseases: A proposal based on the disorders of the three-layered defense systems, barrier, innate immunity and acquired immunity. Journal of Dermatological Science. 2014;**76**(2):81-89

[5] Tabassum N, Hamdani M. Plants used to treat skin diseases. Pharmacognosy Reviews. 2014;**8**(15):52

[6] Torre K, Shahriari M. Clinical trials in dermatology. International Journal of Women's Dermatology. 2017;**3**(3):180-183

[7] Eaglstein WH, Cash KR, Corcoran GR. Challenges encountered in dermatologic drug development. Actas Dermo-Sifiliográficas. 2009;**100**:86-91

[8] Rozmer Z, Perjési P. Naturally occurring chalcones and their biological activities. Phytochemistry Reviews. 2014;**15**(1):87-120

[9] Fathiazad F, Afshar J. An investigation on methylation methods of hesperidin. DARU Journal of Pharmaceutical Sciences. 2004;**12**(2):67-70

[10] Zhuang C, Zhang W, Sheng C, Zhang W, Xing C, Miao Z. Chalcone: A privileged structure in medicinal chemistry. Chemical Reviews. 2017;**117**(12):7762-7810

[11] Ohkatsu Y, Satoh T. Antioxidant and photo-antioxidant activities of chalcone derivatives. Journal of the Japan Petroleum Institute. 2008;**51**(5):298-308

[12] Rashid H, Xu Y, Nasir A, Muhammad Y, Lisheng W. Promising anti-inflammatory effects of chalcones via inhibition of cyclooxygenase, prostaglandin E2, inducible NO synthase and nuclear factor κb activities. Bioorganic Chemistry. 2019;**87**(2018):335-365

[13] Safdar MH, Hasan H, Afzal S, Hussain Z. Exploring promising immunomodulatory potential of natural and synthetic 1,3-diphenyl-2-propen-1-one analogs: A review of mechanistic insight. Mini-reviews. Medicinal Chemistry. 2017;**18**(12):1047-1063

[14] Mojzis J, Varinska L, Mojzisova G, Kostova I, Mirossay L. Antiangiogenic effects of flavonoids and chalcones. Pharmacological Research. 2008;**57**(4):259-265

[15] Mahapatra DK, Bharti SK, Asati V. Chalcone scaffolds as anti-infective agents: Structural and molecular target perspectives. European Journal of Medicinal Chemistry. 2015;**101**:496-524

[16] Zhou B, Xing C. Diverse molecular targets for chalcones with varied bioactivities. Medicinal Chemistry. 2015;**5**(8):388-404

[17] Soter NA. Acute effects of ultraviolet radiation on the skin. In: Lim HW, Soter NA, editors. Clinical Photomedicine. New York: Marcel Dekker, Inc.; 1993. pp. 75-93

[18] Young AR. Chromophores in human skin. Physics in Medicine and Biology. 1997;**42**:789-802

[19] Hart PH, Norval M, Byrne SN, Rhodes LE. Exposure to ultraviolet radiation in the modulation of human diseases. Annual Review of Pathology: Mechanisms of Disease. 2018;**14**(1):55-81

[20] Girard PM, Francesconi S, Pozzebon M, Graindorge D, Rochette P, Drouin R, et al. UVA-induced damage to DNA and proteins: Direct versus indirect photochemical processes. Journal of Physics Conference Series. 2011;**261**(1):1-10

[21] de Jager TL, Cockrell AE, Du Plessis SS. Ultraviolet light induced generation of reactive oxygen species. In: Ahmad SI, editor. Ultraviolet Light in Human Health, Diseases and Environment. Gewerbestrasse: Springer International Publishing AG; 2017. pp. 15-23

[22] Hruza LL, Pentland AP. Mechanisms of UV-induced inflammation. The Journal of Investigative Dermatology. 1993;**100**(1):35S-41S

[23] Tejima T, Nakamura K, Hattori M, Masuda S, Imokawa G, Takaishi N. Novel chalcone derivatives and ultraviolet absorbers comprising the same. United State 4,584,190; 1986

[24] Xue Y, Mou J, Liu Y, Gong X, Yang Y, An L. An ab initio simulation of the UV/visible spectra of substituted chalcones. Central European Journal of Chemistry. 2010;**8**(4):928-936

[25] Shin D, Song D, Jung K, Moon J. Photochemical transformation of chalcone derivatives. Journal of Photoscience. 2001;**8**(1):9-12

[26] Sisa M, Bonnet SL, Ferreira D, Van Der Westhuizen JH. Photochemistry of flavonoids. Molecules. 2010;**15**(8):5196-5245

[27] Kostyuk V, Potapovich A, Albuhaydar AR, Mayer W, De Luca C, Korkina L. Natural substances for prevention of skin photoaging: Screening systems in the development of sunscreen and rejuvenation cosmetics. Rejuvenation Research. 2018;**21**(2):91-101

[28] Martinez RM, Pinho-Ribeiro FA, Steffen VS, Caviglione CV, Fattori V, Bussmann AJC, et al. *Trans* -Chalcone, a flavonoid precursor, inhibits UV-induced skin inflammation and oxidative stress in mice by targeting NADPH oxidase and cytokine production. Photochemical & Photobiological Sciences. 2017;**16**(7):1162-1173

[29] Martinez RM, Pinho-Ribeiro FA, Vale DL, Steffen VS, Vicentini FTMC, Vignoli JA, et al. *Trans*-chalcone added in topical formulation inhibits skin inflammation and oxidative stress in a model of ultraviolet B radiation skin damage in hairless mice. Journal of Photochemistry and Photobiology B: Biology. 2017;**171**:139-146

[30] Lahorkar PGR, Vaidya AA, Chavan MV, Gadgil VR. A photo-protective personal care composition. Nederland; WO 2014/191143; 2014

[31] tom DK, Immeyer J, Wolber R, Kolbe L. Anti-inflammatory properties of Licochalcone a from Glycyrrhiza inflata on various human skin cells. Journal of the American Academy of Dermatology. 2005;**52**(3):P1055

[32] Kolbe L, Immeyer J, Batzer J, Wensorra U, Dieck KT, Mundt C, et al. Anti-inflammatory efficacy of Licochalcone A: Correlation of clinical potency and in vitro effects. Archives of Dermatological Research. 2006;**298**(1):23-30

[33] Kühnl J, Roggenkamp D, Gehrke SA, Stäb F, Wenck H, Kolbe L, et al. Licochalcone A activates Nrf2 in vitro and contributes to licorice extract-induced lowered cutaneous oxidative stress in vivo. Experimental Dermatology. 2015;**24**(1):42-47

[34] Immeyer J, Batzer J, Wolber R, Kolbe L. Anti-irritative efficacy of licochalcone A-containing formulations on razor- and UV-induced skin irritation. Journal of the American Academy of Dermatology. 2005;**52**(3):P1007

[35] Petrova A, Davids LM, Rautenbach F, Marnewick JL. Photoprotection by honeybush extracts, hesperidin and mangiferin against UVB-induced skin damage in SKH-1 mice. Journal of Photochemistry and Photobiology B: Biology. 2011;**103**(2):126-139

[36] Martinez RM, Pinho-Ribeiro FA, Steffen VS, Caviglione CV, Pala D, Baracat MM, et al. Topical formulation containing hesperidin methyl chalcone inhibits skin oxidative stress and inflammation induced by ultraviolet B irradiation. Photochemical & Photobiological Sciences. 2016;**15**(4):554-563

[37] Kuwano T, Watanabe M, Kagawa D, Murase T. Hydrolyzed methylhesperidin induces antioxidant enzyme expression via the Nrf2-ARE pathway in normal human epidermal keratinocytes. Journal of Agricultural and Food Chemistry. 2015;**63**(36):7937-7944

[38] Leu SJ, Lin YP, Lin RD, Wen CL, Cheng KT, Hsu FL, et al. Phenolic constituents of *Malus doumeri* var. *formosana* in the field of skin care. Biological and Pharmaceutical Bulletin. 2006;**29**(4):740-745

[39] Oresajo C, Stephens T, Hino PD, Law RM, Yatskayer M, Foltis P, et al. Protective effects of a topical antioxidant mixture containing vitamin C, ferulic acid, and phloretin against ultraviolet-induced photodamage in human skin. Journal of Cosmetic Dermatology. 2008;**7**:290-297

[40] Brenner M, Hearing VJ. The protective role of melanin against UV damage in human skin. Photochemistry and Photobiology. 2008;**84**(3):539-549

[41] D'Mello SAN, Finlay GJ, Baguley BC, Askarian-Amiri ME. Signaling pathways in melanogenesis. International Journal of Molecular Sciences. 2016;**17**(1141):1-18

[42] Videira IF dos S, Moura DFL, Magina S. Mechanisms regulating melanogenesis. Anais Brasileiros de Dermatologia. 2013;**88**(1):76-83

[43] Rojekar MV, Sawant SD. A short review of pigmentation disorders in systemic diseases. Pigmentary Disorders. 2015;**2**(4):2-4

[44] Nicolaidou E, Katsambas AD. Pigmentation disorders: Hyperpigmentation and hypopigmentation. Clinics in Dermatology. 2014;**32**(1):66-72

[45] Schalka S. New data on hyperpigmentation disorders. Journal of the European Academy of Dermatology and Venereology. 2017;**31**:18-21

[46] Filoni A, Mariano M, Cameli N. Melasma: How hormones can modulate skin pigmentation. Journal of Cosmetic Dermatology. 2018;**2019**:1-6

[47] Lee A. An updated review of melasma pathogenesis. Dermatologica Sinica. 2014;**32**(4):233-239

[48] Kostopoulou I, Detsi A. Recent developments on tyrosinase inhibitors based on the chalcone and aurone scaffolds. Current Enzyme Inhibition. 2018;**14**:3-17

[49] Mann T, Gerwat W, Batzer J, Eggers K, Scherner C, Stäb F, et al. Inhibition of human tyrosinase requires molecular motifs distinctively different from mushroom tyrosinase. The Journal of Investigative Dermatology. 2018;**138**(7):1601-1608

[50] Kim BH, Park KC, Park JH, Lee CG, Ye SK, Park JY. Inhibition of tyrosinase activity and melanin production by the chalcone derivative 1-(2-cyclohexylmethoxy-6-hydroxy-phenyl)-3-(4-hydroxymethyl-phenyl)-propenone. Biochemical and Biophysical Research Communications. 2016;**480**(4):648-654

[51] Kim B, Hong S, Ye S, Park J. Evaluation and optimization of the anti-melanogenic activity of 1-(2-cyclohexylmethoxy-6-hydroxy-phenyl)-3-(4-hydroxymethyl-phenyl)-propenone derivatives. Molecules. 2019;**24**(1372):1-15

[52] Toossi P, Esmaili-Azad M, Saeedi M. Evaluation of licorice efficacy on melasma. Iranian Journal of Dermatology. 2013;**16**(65):118-119

[53] Meymandi SS, Shanehsaz SM, Dogaheh MA, Jahani Y. Efficacy of licorice extracts in the treatment of melasma: Randomized, double-blinded and placebo controlled clinical trial. Journal of Cosmetic Dermatology. 2016;**7**(1):1-9

[54] Nerya O, Vaya J, Musa R, Izrael S, Ben-Arie R, Tamir S. Glabrene and isoliquiritigenin as tyrosinase inhibitors from licorice roots. Journal of Agricultural and Food Chemistry. 2003;**51**(5):1201-1207

[55] Khatib S, Nerya O, Musa R, Shmuel M, Tamir S, Vaya J. Chalcones as potent tyrosinase inhibitors: The importance of a 2,4-substituted resorcinol moiety. Bioorganic & Medicinal Chemistry. 2005;**13**(2):433-441

[56] Lin YP, Hsu FL, Chen CS, Chern JW, Lee MH. Constituents from the Formosan apple reduce tyrosinase activity in human epidermal melanocytes. Phytochemistry. 2007;**68**(8):1189-1199

[57] Rahman R, Hasija Y. Exploring vitiligo susceptibility and management: A brief review. Biomedical Dermatology. 2018;**2**(1):1-13

[58] Spritz RA, Andersen GHL. Genetics of vitiligo. Dermatologic Clinics. 2017;**35**(2):245-255

[59] Wang Y, Li S, Li C. Perspectives of new advances in the pathogenesis of vitiligo: From oxidative stress to autoimmunity. Medical Science Monitor. 2019;**25**:1017-1023

[60] Choi D, Isedeh P, Hamzavi I. Vitiligo: A review of the pathogenesis. Journal of the Egyptian Women's Dermatologic Society. 2014;**11**(3):145-158

[61] Dwivedi M, Laddha N, Begum R. Viral causes of vitiligo: A new perspective for vitiligo pathogenesis. Viral Immunology. 2018;**2**(8):1-4

[62] Lazarova R, Hristakieva E, Lazarov N, Shani J. Vitiligo-related neuropeptides in nerve fibers of the skin. Archives of Physiology and Biochemistry. 2000;**108**(3):262-267

[63] Niu C, Yin L, Nie LF, Dou J, Zhao JY, Li G, et al. Synthesis and bioactivity of novel isoxazole chalcone derivatives on tyrosinase and melanin synthesis in murine B16 cells for the treatment of vitiligo. Bioorganic & Medicinal Chemistry. 2016;**24**(21):5440-5448

[64] Ratri SA. Sintesis dan Uji Aktivitas Senyawa Turunan Kalkon Sebagai Tabir Surya Serapan Lebar dan Agen Filter Fototerapi. Thesis. Faculty of Mathematics and Natural Science. Yogyakarta: Universitas Gadjah Mada; 2018

[65] Tuerxuntayi A, Liu Y, Tulake A, Kabas M, Eblimit A, Aisa HA. Kaliziri extract upregulates tyrosinase, TRP-1, TRP-2 and MITF expression in murine B16 melanoma cells. BMC Complementary and Alternative Medicine. 2014;**14**(166):1-9

[66] Niu C, Tuerxuntayi A, Li G, Kabas M, Dong CZ, Aisa HA. Design, synthesis and bioactivity of chalcones and its analogues. Chinese Chemical Letters. 2017;**28**(7):1533-1538

[67] Burza S, Croft SL, Boelaert M. Leishmaniasis. Lancet. 2018;**392**(10151):951-970

[68] Colmenares M, Kar S, Goldsmith-Pestana K, McMahon-Pratt D. Mechanisms of pathogenesis: Differences amongst Leishmania species. Transactions of the Royal Society of Tropical Medicine and Hygiene. 2004;**96**(S1):S1/3-S1/7

[69] David CV, Craft N. Cutaneous and mucocutaneous leishmaniasis. Dermatologic Therapy. 2009;**22**(6):491-502

[70] Akhoundi M, Kuhls K, Cannet A, Votýpka J, Marty P, Delaunay P, et al. A historical overview of the classification, evolution, and dispersion of leishmania parasites and sandflies. PLoS Neglected Tropical Diseases. 2016;**10**(3):1-40

[71] Killick-Kendrick R. The biology and control of Phlebotomine sand flies. Clinics in Dermatology. 1999;**17**(3):279-289

[72] Vannier-Santos MA, Martiny A, de SW. Cell biology of Leishmania spp.: Invading and evading. Current Pharmaceutical Design. 2002;**8**(4):297-318

[73] Kaye P, Scott P. Leishmaniasis: Complexity at the host-pathogen interface. Nature Reviews. Microbiology. 2011;**9**(8):604-615

[74] Zhai L, Blom J, Chen M, Christensen SB, Kharazmi A. The antileishmanial agent licochalcone A interferes with the function of parasite mitochondria. Antimicrobial Agents and Chemotherapy. 1995;**39**(12):2742-2748

[75] Chen M, Zhai L, Christensen SB, Theander TG, Kharazmi A. Inhibition of fumarate reductase in Leishmania major and l. donovani by chalcones. Antimicrobial Agents and Chemotherapy. 2001;**45**(7):2023-2029

[76] Ogungbe IV, Erwin WR, Setzer WN. Antileishmanial phytochemical phenolics: Molecular docking to potential protein targets. Journal of Molecular Graphics & Modelling. 2014;**48**:105-117

[77] Gomes MN, Alcântara LM, Neves BJ, Melo-Filho CC, Freitas-Junior LH, Moraes CB, et al. Computer-aided discovery of two novel chalcone-like compounds active and selective against Leishmania infantum. Bioorganic & Medicinal Chemistry Letters. 2017;**27**(11):2459-2464

[78] Boeck P, Falcão CAB, Leal PC, Yunes RA, Filho VC, Torres-Santos EC, et al. Synthesis of chalcone analogues with increased antileishmanial activity. Bioorganic & Medicinal Chemistry. 2006;**14**(5):1538-1545

[79] de M c B, Argenta DF, Melchiades G de L, Cordeiro MNS, Tonini ML, Moraes MH, et al. Nanoemulsions containing a synthetic chalcone as an alternative for treating cutaneous leshmaniasis: Optimization using a full factorial design. International Journal of Nanomedicine. 2015;**10**:5529-5542

[80] Miranda-Sapla MM, Tomiotto-Pellissier F, Assolini JP, Carloto ACM, Bortoleti BT d S, Gonçalves MD, et al. *trans*-Chalcone modulates Leishmania amazonensis infection in vitro by Nrf2 overexpression affecting iron

availability. European Journal of Pharmacology. 2019;**853**:275-288

[81] Sandjo LP, de Moraes MH, Kuete V, Kamdoum BC, Ngadjui BT, Steindel M. Individual and combined antiparasitic effect of six plant metabolites against Leishmania amazonensis and Trypanosoma cruzi. Bioorganic & Medicinal Chemistry Letters. 2016;**26**(7):1772-1775

[82] de Jesus Sousa-Batista A, Pacienza-Lima W, Arruda-Costa N, CAB F, Ré MI, Rossi-Bergmann B. Depot subcutaneous injection with chalcone CH8-loaded poly(lactic-co-glycolic acid) microspheres as a single-dose treatment of cutaneous leishmaniasis. Antimicrobial Agents and Chemotherapy. 2018;**62**(3):1-11

[83] Piñero J, Temporal RM, Silva-Gonçalves AJ, Jiménez IA, Bazzocchi IL, Oliva A, et al. New administration model of *trans*-chalcone biodegradable polymers for the treatment of experimental leishmaniasis. Acta Tropica. 2006;**98**(1):59-65

[84] Andrighetti-Fröhner CR, de Oliveira KN, Gaspar-Silva D, Pacheco LK, Joussef AC, Steindel M, et al. Synthesis, biological evaluation and SAR of sulfonamide 4-methoxychalcone derivatives with potential antileishmanial activity. European Journal of Medicinal Chemistry. 2009;**44**(2):755-763

[85] Bello ML, Chiaradia LD, Dias LRS, Pacheco LK, Stumpf TR, Mascarello A, et al. Trimethoxy-chalcone derivatives inhibit growth of Leishmania braziliensis: Synthesis, biological evaluation, molecular modeling and structure-activity relationship (SAR). Bioorganic & Medicinal Chemistry. 2011;**19**(16):5046-5052

[86] Roussaki M, Hall B, Lima SC, Da Silva AC, Wilkinson S, Detsi A. Synthesis and anti-parasitic activity of a novel quinolinone-chalcone series. Bioorganic & Medicinal Chemistry Letters. 2013;**23**(23):6436-6441

[87] Ramírez-Prada J, Robledo SM, Vélez ID, Crespo MP, Quiroga J, Abonia R, et al. Synthesis of novel quinoline–based 4,5–dihydro–1H–pyrazoles as potential anticancer, antifungal, antibacterial and antiprotozoal agents. European Journal of Medicinal Chemistry. 2017;**131**:237-254

[88] de Mello MVP, de Abrahim-Vieira BA, TFS D, de Jesus JB, de Sousa ACC, Rodrigues CR, et al. A comprehensive review of chalcone derivatives as antileishmanial agents. European Journal of Medicinal Chemistry. 2018;**150**:920-929

[89] Williams HC, Dellavalle RP, Garner S. Acne vulgaris. Lancet. 2012;**379**(9813):361-372

[90] Thiboutot D, Gollnick H, Bettoli V, Dréno B, Kang S, Leyden JJ, et al. New insights into the management of acne: An update from the global alliance to improve outcomes in acne group. Journal of the American Academy of Dermatology. 2009;**60**(5):S1-S50

[91] Wang Y, Kuo S, Shu M, Yu J, Huang S, Dai A, et al. Staphylococcus epidermidis in the human skin microbiome mediates fermentation to inhibit the growth of Propionibacterium acnes: Implications of probiotics in acne vulgaris. Applied Microbiology and Biotechnology. 2014;**98**(1):411-424

[92] Angelova-Fischer I, Rippke F, Fischer TW, Neufang G, Zillikens D. A double-blind, randomized, vehicle-controlled efficacy assessment study of a skin care formulation for improvement of mild to moderately severe acne. Journal of the European Academy of Dermatology and Venereology. 2013;**27**:6-11

[93] Chularojanamontri L, Tuchinda P, Kulthanan K, Varothai S, Winayanuwattikun W. A double-blinded, randomized, vehicle-controlled study to access skin tolerability and efficacy of an anti-inflammatory moisturizer in treatment of acne with 0.1% adapalene gel. The Journal of Dermatological Treatment. 2016;**27**(2):140-145

[94] Kantikosum K, Chongpison Y, Chottawornsak N, Asawanonda P. The efficacy of glycolic acid, salicylic acid, gluconolactone, and licochalcone A combined with 0.1% adapalene vs adapalene monotherapy in mild-to-moderate acne vulgaris: A double-blinded within-person comparative study. Clinical, Cosmetic and Investigational Dermatology. 2019;**12**:151-161

[95] Yang G, Lee HE, Yeon SH, Kang HC, Cho YY, Lee HS, et al. Licochalcone A attenuates acne symptoms mediated by suppression of NLRP3 inflammasome. Phytotherapy Research. 2018;**32**(12):2551-2559

[96] Kum H, Roh KB, Shin S, Jung K, Park D, Jung E. Evaluation of anti-acne properties of phloretin in vitro and in vivo. International Journal of Cosmetic Science. 2016;**38**(1):85-92

[97] Cheon D, Kim J, Jeon D, Shin HC, Kim Y. Target proteins of phloretin for its anti-inflammatory and antibacterial activities against propionibacterium acnes-induced skin infection. Molecules. 2019;**24**(1319):1-14

[98] Yamaguchi N, Satoh-Yamaguchi K, Ono M. In vitro evaluation of antibacterial, anticollagenase, and antioxidant activities of hop components (Humulus lupulus) addressing acne vulgaris. Phytomedicine. 2009;**16**(4):369-376

[99] Song MS, Shim JS, Gwon SH, Lee CW, Kim HS, Hwang J-K. Antibacterial activity of panduratin a and isopanduratin a isolated from Kaempferia pandurata Roxb. against acne-causing microorganisms. Food Science and Biotechnology. 2008;**17**(6):1357-1360

[100] Takahashi T, Kokubo R, Sakaino M. Antimicrobial activities of eucalyptus leaf extracts and flavonoids from Eucalyptus maculata. Letters in Applied Microbiology. 2004;**39**(1):60-64

[101] Saliou C, Boddupalli S, Mahmood K, Anthonavage M, Huang K. Compositions containing chalcones and use thereof. United States; US 7,781,410 B2; 2010

[102] Wakelin S. Seborrhoeic dermatitis. In: Griffiths CEM, Barker J, Bleiker T, Chalmers R, Creamer D, editors. Rook's Textbook of Dermatology. 9th ed. Chichester: John Wiley & Sons, Ltd; 2016. pp. 40.1-40.8

[103] Szöllősi AG, Oláh A, Bíró T, Tóth BI. Recent advances in the endocrinology of the sebaceous gland. Dermatoendocrinol. 2018;**9**(1):e1361576

[104] Rembiesa J, Ruzgas T, Engblom J, Holefors A. The impact of pollution on skin and proper efficacy testing for anti-pollution claims. Cosmetics. 2018;**5**(1):4

[105] Ro BI, Dawson TL. The role of sebaceous gland activity and scalp microfloral metabolism in the etiology of seborrheic dermatitis and dandruff. The Journal of Investigative Dermatology. Symposium Proceedings. 2005;**10**(3):194-197

[106] Ramsden CE, Domenichiello AF, Yuan Z-X, Sapio MR, Keyes GS, Mishra SK, et al. A systems approach for discovering linoleic acid derivatives that potentially mediate pain and itch. Science Signaling. 2017;**10**(493):eaal5241

[107] Thomas DS, Ingham E, Bojar RA, Holland KT. In vitro modulation of

human keratinocyte pro- and anti-inflammatory cytokine production by the capsule of Malassezia species. FEMS Immunology and Medical Microbiology. 2008;**54**(2):203-214

[108] Gaitanis G, Magiatis P, Hantschke M, Bassukas ID, Velegraki A. The Malassezia genus in skin and systemic diseases. Clinical Microbiology Reviews. 2012;**25**(1):106-141

[109] Schweiger D, Rippke F, Drescher P, Braren S, Lüttke J, Filbry A, et al. Highly efficient rinse-off/leave-on scalp care treatments to reduce moderate to severe dandruff. Journal of Cosmetics, Dermatological Sciences and Applications. 2013;**03**(01):46-55

[110] Wananukul S, Chatproedprai S, Charutragulchai W. Randomized, double-blind, split-side comparison study of moisturizer containing licochalcone vs. 1% hydrocortisone in the treatment of infantile seborrhoeic dermatitis. Journal of the European Academy of Dermatology and Venereology. 2012;**26**(7):894-897

[111] Zeng J, Luo S, Huang Y, Lu Q. Critical role of environmental factors in the pathogenesis of psoriasis. The Journal of Dermatology. 2017;**44**(8):863-872

[112] Chandra A, Ray A, Senapati S, Chatterjee R. Genetic and epigenetic basis of psoriasis pathogenesis. Molecular Immunology. 2015;**64**(2):313-323

[113] Lee EB, Amin M, Bhutani T, Wu JJ. Emerging therapies in psoriasis: A systematic review. Cutis. 2018;**101**(3):5-9

[114] Pfaff CM, Marquardt Y, Fietkau K, Baron JM, Lüscher B. The psoriasis-associated IL-17A induces and cooperates with IL-36 cytokines to control keratinocyte differentiation and function. Scientific Reports. 2017;**7**(1):1-13

[115] Kofoed K, Skov L, Zachariae C. New drugs and treatment targets in psoriasis. Acta Dermato-Venereologica. 2015;**95**(2):133-139

[116] Wu Y, Chen X, Ge X, Xia H, Wang Y, Su S, et al. Isoliquiritigenin prevents the progression of psoriasis-like symptoms by inhibiting NF-κB and proinflammatory cytokines. Journal of Molecular Medicine. 2016;**94**(2):195-206

[117] Taylor EJ, Evans FJ. Chalcone plant extracts for use in therapy. England, EP 0998939A1; 2000. pp. 1-14

[118] Pinz S, Unser S, Brueggemann S, Besl E, Al-Rifai N, Petkes H, et al. The synthetic α-bromo-2′,3,4,4′-tetramethoxychalcone (α-Br-TMC) inhibits the JAK/STAT signaling pathway. PLoS One. 2014;**9**(3):1-13

[119] Jobst B, Weigl J, Michl C, Vivarelli F, Pinz S, Amslinger S, et al. Inhibition of interleukin-3- and interferon- α-induced JAK/STAT signaling by the synthetic α-X-2′,3,4,4′-tetramethoxychalcones α-Br-TMC and α-CF3-TMC. Biological Chemistry. 2016;**397**(11):1187-1204

[120] Rücker H, Al-Rifai N, Rascle A, Gottfried E, Brodziak-Jarosz L, Gerhäuser C, et al. Enhancing the anti-inflammatory activity of chalcones by tuning the Michael acceptor site. Physical Chemistry Chemical Physics. 2015;**13**(10):3040-3047

[121] Weidinger S, Novak N. Atopic dermatitis. Lancet. 2016;**387**:1109-1122

[122] Brown SJ, McLean WHI. Eczema genetics: Current state of knowledge and future goals. The Journal of Investigative Dermatology. 2009;**129**(3):543-552

[123] Egawa G, Weninger W. Pathogenesis of atopic dermatitis: A short review. Cogent Biology. 2015;**1**(1):1-10

[124] Bonamonte D, Filoni A, Vestita M, Romita P, Foti C, Angelini G. The role of the environmental risk factors in the pathogenesis and clinical outcome of atopic dermatitis. BioMed Research International. 2019;**2019**:1-11

[125] Udompataikul M, Srisatwaja W. Comparative trial of moisturizer containing licochalcone A vs. hydrocortisone lotion in the treatment of childhood atopic dermatitis: A pilot study. Journal of the European Academy of Dermatology and Venereology. 2011;**25**(6):660-665

[126] Angelova-Fischer I, Neufang G, Jung K, Fischer TW, Zillikens D. A randomized, investigator-blinded efficacy assessment study of stand-alone emollient use in mild to moderately severe atopic dermatitis flares. Journal of the European Academy of Dermatology and Venereology. 2014;**28**(3):9-15

[127] Boonchai W, Varothai S, Winayanuwattikun W, Phaitoonvatanakij S, Chaweekulrat P, Kasemsarn P. Randomized investigator-blinded comparative study of moisturizer containing 4-t-butylcyclohexanol and licochalcone A versus 0.02% triamcinolone acetonide cream in facial dermatitis. Journal of Cosmetic Dermatology. 2018;**17**(6):1130-1135

[128] Angelova-Fischer I, Rippke F, Richter D, Filbry A, Arrowitz C, Weber T, et al. Stand-alone emollient treatment reduces flares after discontinuation of topical steroid treatment in atopic dermatitis: A double-blind, randomized, vehicle-controlled, left-right comparison study. Acta Dermato-Venereologica. 2018;**98**(5):517-523

[129] Yu H, Li H, Li Y, Li M, Chen G. Effect of isoliquiritigenin for the treatment of atopic dermatitis-like skin lesions in mice. Archives of Dermatological Research. 2017;**309**(10):805-813

[130] Choi BM, Oh GS, Lee JW, Mok JY, Kim DK, Il JS, et al. Prenylated chalcone from Sophora flavescens suppresses Th2 chemokine expression induced by cytokines via heme oxygenase-1 in human keratinocytes. Archives of Pharmacal Research. 2010;**33**(5):753-760

[131] Steinhoff M, Schauber J, Leyden JJ. New insights into rosacea pathophysiology: A review of recent findings. Journal of the American Academy of Dermatology. 2013;**69**(6):15-26

[132] Reinholz M, Tietze JK, Kilian K, Schaller M, Schöfer H, Lehmann P, et al. Rosazea - S1 Guideline. Journal of the German Society of Dermatology. 2013;**11**(8):768-779

[133] Anzengruber F, Czernielewski J, Conrad C, Feldmeyer L, Yawalkar N, Häusermann P, et al. Swiss S1 guideline for the treatment of rosacea. Journal of the European Academy of Dermatology and Venereology. 2017;**31**(11):1775-1791

[134] Gilbert D, Hartmane I, Ancupane I, Kolontaja I, Princeva O, Rozniece K, et al. Efficacy and tolerance of a topical skin care regimen as an adjunct to treatment of facial rosacea. Cosmetic Dermatology. 2008;**21**(9):501-504

[135] Pinho-Ribeiro FA, Hohmann MSN, Borghi SM, Zarpelon AC, Guazelli CFS, Manchope MF, et al. Protective effects of the flavonoid hesperidin methyl chalcone in inflammation and pain in mice: Role of TRPV1, oxidative stress, cytokines and NF-κB. Chemico-Biological Interactions. 2015;**228**:88-99

[136] Weber T, Schoelermann A, Buerger A, Rizer R, Stephens TJ.

Tolerance and efficacy of a skin care regimen containing Licochalcone A for adults with erythematous rosacea and facial redness. Journal of the American Academy of Dermatology. 2005;**52**(3):P95

[137] Broniarczyk-Dyła G, Prusińska-Bratoś M, Kmieć ML. Assessment of the influence of licochalcone on selected functional skin parameters in patients with impaired vasomotor disorders and rosacea. Postepy Dermatologii i Alergologii. 2011;**28**(4):241-257

[138] Gafner F, Schmid D, Lozza J, Zülli F. Tetra-carboxy-methyl-naringenin- chalcone, a new active to treat rosacea. Household and Personal Care Today. 2013;**8**(5):14-16

Chapter 8

Maxillofacial Defects: Impact on Psychology and Esthetics

Poonam Prakash, Rahul Bahri and S.K. Bhandari

Abstract

Maxillofacial defects arising due to developmental anomalies, trauma or ablative cancer surgeries pose a challenge to an individual due to alteration in form, function and esthetics. Face is considered to be a reflection of one's personality and existence. Any alteration in facial structures or symmetry alters the esthetics of an individual. This may have a deep psychological impact on the patient affecting self-confidence, self-worth and ability to interact among peers. Maxillofacial Prosthodontics is a specialty that deals with rehabilitation of missing or deformed structures of orofacial region to achieve normalcy as much as practically feasible. A multidisciplinary approach is required to evaluate the psychological status, understand the impact of defect altering esthetics and mental make-up of the patient and follow an organized approach to alleviate the impact of maxillofacial defects in overall life of an individual. This chapter highlights the impact of maxillofacial defect on esthetics of an individual and psychological impact of the same.

Keywords: Psychological evaluation, maxillofacial defect, esthetics

1. Introduction

"It is the God given right of every human being to appear human" [1].

Beauty is the quality or an aggregate of qualities in a person/ thing that gives pleasure to the senses or pleasurable exalts to the mind or spirit. Esthetics is the branch of philosophy dealing with beauty. Body image is considered as "the lifelong anchor for self-awareness" as it is closely related to sense of adequacy and competence [2]. A symmetrical, proportionate facial appearance with equal horizontal thirds and vertical fifths, gives an appearance of esthetically beautiful face. Various proportions like golden proportion exist in nature which gives a perception of dynamic symmetry. Lombardi proposed various principles for esthetics based on laws of nature [3].

Mouth is the most dynamic component of face which provides a unity with variety. **According to Sigmund Freud's theory 'Oral phase" as earliest expressions of "self". Since very early age, mouth is considered as an area of gratification and security. Orofacial region is the reflection of personality, image and primary mode of self- expression for growing young adults and a gateway for proper diet and nutrition in elderly.**

Any defect in the maxillofacial region leads to loss of form, function and esthetics. It compromises the integrity of craniofacial region and thus poses a deeper impact on an individual altering his personal and social acceptability to a major extent. Rehabilitation of these defects surgically or Prosthodontically restores the function, esthetics to near normalcy and elevates the sense of 'incomplete' to an extent but deep seated insecurities and psychological impact needs to be dealt with, to ensure complete rehabilitation of patient and acceptance in society.

Defects in the maxillofacial region may be intraoral including maxillary defect, mandibular defect affecting continuity of mandible, velopharyngeal defects or defects of soft palate and extraoral defects like residual ocular, auricular, orbital, cranial, nasal or combination defects.

Due to variations in the site and size of defect, the impact of defect on various spheres of an individual's life, the psychological make-up and ability to cope up, esthetic expectations titrated with the realistic rehabilitation options makes maxillofacial rehabilitation a challenging task.

2. Discussion

2.1 Loss and grief in maxillofacial defect

Loss is defined as 'a state of being deprived of or being without something one has had and valued' [4]. According to Peretz, loss may be in form of loss of significant person, Loss of a part of the self, Loss of material object or Developmental loss. Loss of part of human body especially in the esthetic regions like craniofacial region, puts a deep psychological stress on an individual. From the diagnosis of pathology itself to the surgical trauma, followed by healing phase, rehabilitation and finally maintenance and follow up puts individual into a progressive cycle of grief at different stages. The reaction to grief is an adaptive function 'to assure group cohesiveness in species where a social form of existence is necessary for survival' [4]. It follows a pattern of shock and denial characterized by signs of depression and suicidal tendencies in some patients. This is followed by a phase of guilt, anger and a search to find ways to discharge emotional pain. Eventually, adjustment, acceptance and growth takes over where the patient acceptance to loss, tries to make healthy adjustments and formulates new life patterns [4]. A cycle of loss, grief and reintegration must be completed by the patient and understood by the prosthodontist. The role of a specialist is to empathize with the patient and evaluate the need for referral to a psychotherapist.

3. Esthetic impact of maxillofacial defect

Face is the individuality of a person, it is a medium of conveying various emotions, expressions and governs the personality of an individual. Based on age, gender, occupation, social interactions and interpersonal relations, different individuals weigh their appearance differently. But a disfigured, asymmetric face grabs first attention. **Prior to involving the orofacial region, though affected but its integrity is maintained. Post maxillofacial surgery, the resected site leads to loss of continuity and causes an altered disfigured appearance. This is one of the major factors contributing to distress, shame and psychosocial burden for the patient.** Post-surgical depression, hampers healing and makes patient prone to infections [5]. Surgical reconstruction or Prosthodontic rehabilitation attempts to restore the function and esthetics to as near natural as possible with its own limitations and constraints. **The constraints includes anatomical limitations which may limit the extension of the prosthesis, material based on area of application i.e. intraoral or extraoral, patients mental state and ability to accept.**

4. Psychological impact of maxillofacial defect

Maxillofacial defects may be congenital, developmental or acquired. **Acquired defects produce a deeper impact as the individual is not born with the same and is affected later in life, resulting in drastic change in appearance, post-surgical or after Prosthodontic intervention.** They alter the integrity of craniofacial region leading to disruption in normal functioning, maintenance of form and esthetic appearance of an individual affecting overall psychological status of the patient. **The condition might also alter the individual's financial, spiritual and social status due to irreversible loss. Impaired social status due to physical disfigurement and mental impact. Individuals are also affected spiritually as they question 'Why Me'**. Surgical or Prosthodontic modalities help in rehabilitation of the patient to bridge the gap and help to restore his individuality in the society. The whole process may manifest itself into various kinds of psychological impairment including psychoneurotic, psychotic or personality disorders and the treating clinician must observe for clinical signs hinting towards these [6]. These may include:

Anxiety disorders including panic disorder, obsessive-compulsive disorder (OCD), posttraumatic stress disorder (PTSD), social phobia, and generalized anxiety disorder.

Mood disorders or **affective disorders**, create disturbances in a person's emotional life. Depression, mania, and bipolar disorder are examples of mood disorders.

People with schizophrenia and other psychotic disorders lose contact with reality. Symptoms may include delusions and hallucinations, disorganized thinking and speech, bizarre behavior, a diminished range of emotional responsiveness and social withdrawal.

Personality Disorders are mental illnesses in which one's personality results in personal distress or a significant impairment in social or work functioning.

Cognitive disorders, such as delirium and dementia, involve a significant loss of mental functioning. **Dissociative disorders** involve disturbances in a person's consciousness, memories, identity and perception of the environment [7].

A patient suffering from any mental or behavioral disorder sustains a deeper impact of the disease or the residual maxillofacial defect produced as a result of surgical correction due to altered or reduced tolerance.

5. Psychological classification and interpretation

Evaluation of mental attitude of patients has been an important aspect in clinical case history in Prosthodontics. Various authors like House, Heartwell, Sharry, Gamer et al and others have classified patients based on evaluation of psychological status [8].

This approach may be used for the conventional prosthodontic patient, however in patients with disfigured orofacial region or altered biological function like speech or swallowing it may not hold valid due to the impact being much stronger. In such patients, underlying deep seated emotional distress may get exaggerated. Further, in such an individuals it is critical for the prosthodontist to assess the mental status of the patient, need for referral and modification of treatment modality best suited for the patient.

It becomes paramount that the prosthodontist understand the various psychological diagnoses, ranging from subtle emotional distress to overt psychological disorders, that potentially undermine successful prosthodontic

treatment. For example, if a patient taken up for Prosthodontic rehabilitation, but shows frequent bouts of crying or abrupt temper tantrum is indicative of a depressive mental state or altered psychological condition including panic disorder, obsessive-compulsive disorder (OCD), posttraumatic stress disorder (PTSD), mood disorder or a bipolar state. This may hint the treating prosthodontist for the need of intervention of psychologist or a psychotherapist for better treatment outcome.

6. Management

Management of maxillofacial defects is a multidisciplinary, well planned execution of procedures starting pre-surgery and extending long after rehabilitation is complete. It includes patient awareness, early diagnosis & prompt treatment, role of Healthcare specialist/worker, role of a psychological counselor, family and support groups.

6.1 Patient awareness, early diagnosis & prompt treatment

Awareness, early diagnosis and prompt treatment can help minimize the spread of habit related orofacial cancers, involvement of normal anatomical structures and the extent and size of defect requiring rehabilitation.

Various awareness programs and statutory warnings like increased chances of cancers due to cigarette or tobacco consumption are displayed on television, radio and alongside streets in many Asian countries for awareness of patients about the harmful effects and the dreaded consequences. Also advertisements about treatment options available and life post-rehabilitation are propagated to spread awareness and reduce anxiety among patients.

6.2 Role of Healthcare specialist / worker

Management of patients with maxillofacial defect is a multidisciplinary approach and involves specialists from various branches. Prosthodontic rehabilitation allows replacement of lost structures with artificial substitutes to near normal appearance as possible. For extraoral defects, esthetics is the main concern. Various measures are taken to ensure correct anatomical alignment of the artificial prosthesis, well adaptation, proper retention using retentive aids including anatomical, mechanical or physical means and shade matching. Advancements in technology and introduction of three dimensional scanning techniques using extraoral scanners, reconstructions using DICOM image files, rapid prototyping, and spectrophotometer have enhanced the precision that can be achieved in maxillofacial prosthetics. **For intraoral defect, advancements in material like PEEK and modifications in designs allows to achieve more esthetic results by concealing the metal display and enhancing appearance with function** [9].

To allow patient to accept the treatment and comply with the instructions, it is important that the patient is in psychological state of mind to accept the loss and cope up with the grief. Active listening and empathy are the qualities that the clinician must possess to look past patient's defect and understand the deep seated distress.

Psychotherapeutic techniques help in recognition of the feelings that patient is trying to express and allowing them to express to do so in elaborate form for better understanding and management.

Evaluation of psychological status of patient can be done in form of interviews, rating scales, tests or questionnaires which may be self-administered or by clinician. Some scales to aid in psychological evaluation Becks Depression inventory, Hospital Anxiety and Depression scale, General Health Questionnaire-12/28, ICP, ICF, MMPI and others.

Preoperative care includes understanding that **the diagnosis and breaking the news like cancers puts** the patient into a progressive cycle of loss and grief where they start questioning themselves and their destiny. It is an emotional turmoil for the patient and the family. Letting the patient vent out and come to terms with the reality helps them cope up with the therapy. Patient should be introduced to various survivors groups, AV aids can help in making them understand the progression of the disease and various rehabilitation options available.

The patient and family should be made aware of the emotional reactions and fluctuations they can expect during the postoperative and extended-care phases of treatment. Giving advance warning helps the patient realize that his or her emotional reactions during treatment are normal.

6.3 Postoperative care

Post operatively, the patient is in a vulnerable state as the trauma of surgery added up with the actual loss of part of 'self'. Alteration of facial symmetry, alteration in speech, difficult mastication and esthetic compromise pushes patient into a deep psychological distress. Good communication skills, motivation, psychodynamic therapy, support groups, active listening, empathy can help to improve the mental make-up of the patient and family.

6.4 Extended care & Team work

A rehabilitation specialist, maxillofacial Prosthodontist, reconstructive surgeon are a ray of hope for the patients. A well fitted prosthesis or reconstructive surgical modalities improves patient's condition and enable them to restore form, function and esthetics. Team work including role of psychological counsellor, administration of anti-depressants, nutrition counselling and diet modification to combat the side effects of radiation therapy or dysfunctioning of certain organs like xerostomia, alternate medicine specialists like naturopathy, yoga, mind body exercises can help patient develop a positive mental attitude and improve prognosis of the treatment.

6.5 Case summary

A 42 yrs old male patient reported with a chief complaint of a non-healing ulcer in the palate since 04 months with no facial disfigurement. Patient gave a history of smoking and tobacco chewing since past 08 years. On examination and histopathological investigations, a diagnosis of verrucous carcinoma was arrived at. Treatment plan formulated was surgical resection followed by prosthodontic rehabilitation. On interviewing the patient and history taking, it was felt that the patient was very anxious and concerned about his facial appearance. To check his current state of anxiety, a self administered questionnaire (DASS) was provided and based on the results it was observed that he had moderate-severe anxiety state [10]. His family was addressed and prepared so that they could provide the patient with necessary support needed. When the patient was told about the diagnosis, he was unable to accept it and had bouts of crying. He was counselled and was introduced to a support group with patients who had undergone similar procedure earlier. AV aids were used to motivate the patient and introduce the prosthetic options and their impact on lifestyle. On active listening, it was observed that patient had a financial crunch due to loss of job and that was another reason for his hysteria. He was introduced to local Non-profitable institution who were ready to share the expense for the surgery. Pre-surgery a mock procedure was shown to the patient. During surgery, maxillectomy was performed using Weber-Fergusson incision that resulted in a large intra oral defect and a visible facial scar. His speech, mastication and nutrition was affected. Depressive state was prevalent as observed in his behavior, response, loss of apatite and unwillingness to meet or talk to others. He was kept of psychological counselling sessions for 03 months on weekly basis and a well fitted definitive prosthesis was fabricated. Post 06 months, he was able to adjust to the condition and volunteered to be a member of support group to help others in similar hardships. His psychological state changed from normal to highly anxious and depressive post-surgery and eventually improved to mild-moderate anxious state post rehabilitation.

7. Conclusion

Psychological health and spiritual well-being are integrated and related states. Maxillofacial defects lead to loss of continuity and leave a residual defect intraorally and/or a visible defect extraorally. This leads to altered esthetics and may have a deep psychological impact on well-being of the patient. Health care providers need to understand these conditions and must exhibit empathy while ensuring high quality rehabilitation. A specialist who can learn to actively listen to patients, communicates with them and understand their feelings, will aid in positive results in the management of their patients.

Author details

Poonam Prakash[1*], Rahul Bahri[1] and S.K. Bhandari[2]

1 Prosthodontics and Crown and Bridge, Department of Dental Surgery, Armed Forces Medical College, Pune, India

2 Department of Dental Surgery and Oral Health Sciences, Armed Forces Medical College, Pune, India

*Address all correspondence to: pnmprakash1977@gmail.com

References

[1] Chalian VA, Drane JB, Standish SM. Maxillofacial prosthetics: multidisciplinary practice. Williams & Wilkins Company; 1972.

[2] Allport GW. Becoming: Basic considerations for a psychology of personality. Yale University Press; 1955.

[3] Lombardi RE. The principles of visual perception and their clinical application to denture esthetics. J Prosthet Dent 1973;29(4):358-82.

[4] Taylor TD. Clinical Maxillofacial Prosthetics. Quintessence Publishing Co,Inc. 2000.

[5] Grieder A. Psychologic aspects of prosthodontics. J Prosthet Dent. 1973 Nov 1;30(5)736-44.

[6] GhoneimMM,O'HaraMW.Depression and postoperative complications: an overview. BMC surgery. 2016 Dec 1;16(1):5.

[7] Hickey AJ, Salter M. Prosthodontic and psychological factors in treating patients with congenital and craniofacial defects. J Prosthet Dent 2006;95:392-6.

[8] Deshpande S, Babar G, Jain JK. Different Classification Systems of Complete Denture Patients Based on Mental Attitude: A Review IJOCR 2015; 3(2):28-31

[9] de Caxias F, dos Santos DM, Bannwart LC, et al. Classification, history, and future prospects of maxillofacial prosthesis. Int J Dent 2019;8657619

[10] Parkitny L, McAuley J. The depression anxiety stress scale (DASS). Journal of physiotherapy. 2010;56(3):204.

Section 3

Interpersonal Influences on External and Internal Beauty

Chapter 9

Decoding the Signals of Facial Attractiveness: A Communication Theory Perspective

Daniel Gill

Abstract

The human face is an elaborate communication tool that transmits a large variety of signals such as: identity, gender, ethnicity, age, emotional state, health, and more. Of particular importance is the tendency of human observers to infer social traits (e.g., attractiveness, dominance or trustworthiness) rapidly from faces which, in turn, can lead to a specific action from a wide spectrum of possibilities ranging from mating to violent clashes. Among the social trait signals that are transmitted by the face, the attractiveness signal is outstanding in its robustness against manipulation initiated by the transmitter or interference caused by the physical environment's many aspects. Among these aspects are the robustness of attractiveness to manipulations caused by the physical environment (e.g., viewing distance) or manipulations made by the signaler (e.g., facial movements). To understand what makes a face attractive and the unique role that attractiveness plays as a communication signal, this chapter will use the universal framework of communication systems. Every communication system consists of three key elements: a transmitter, a receiver, and a communication channel. All these three components affect the semantic meaning of every message transmitted in the system and thus shape the outcome following the message reception.

Keywords: action units, communication theory, facial attractiveness, social camouflage, social traits, spatial frequency

1. Introduction

When the Beatles, way back in 1965, sang the following lyrics of their song "I've just seen a face" [1], they summarized, by these naive lines, some of the cognitive processes and social outcomes related to one of the most culturally and socially important parts of the body (second only to the brain)—the face:

> *"I've just seen a face,*
> *I cannot forget the time or place*
> *where we just met.*
> *She's just the girl for me*
> *and I want all the world to see*
> *we have met..."*

Mankind has developed certain skills allowing for fast and reliable processing of facial information. Humans are able to detect faces: that is, the ability to identify and locate all the present faces within the receptive field ("I've just seen a face"). Humans also have the ability to recognize a face: to judge whether a face has been seen before and in case of positive recognition the ability to retrieve semantic data such as name or context of encounter ("I can't forget the time or place..."). People recognize other people by their face and not, for example, by the palm of their hand. The face plays a crucial role in mate quality appraisal ("She's just the girl for me...," asserts Sir Paul McCartney after only seeing her face), and the preference for facial beauty exists from a very young age [2, 3]. A pretty partner is a status symbol [4], that is, a visible indicator of economic or social status ("I want all the world to see we've met...").

Random social encounters can lead to dramatic consequences ranging from trust and romantic relationships to fear and violent clashes. The decision whether to avoid or approach an unknown person therefore involves a quick risk assessment of potential gains and losses. When there is no prior information, such a decision is in many cases based on outward appearance. Humans quickly infer social traits, such as attractiveness, aggressiveness, dominance, and trustworthiness from the physical properties of the bodies and faces of others [5–7]. When it comes to facial appearance, such social inferences can be made after a very short exposure time and with high levels of interpersonal agreement [8–12] consequently affecting social outcomes [13–17]. One of the most studied face-inferred social traits is attractiveness.

The common notion that "Beauty is not judged objectively, but according to the beholder's estimation" dates back at least to the third century BC (Theocritus, The Idyll as cited in [18]). Even Darwin came to the same conclusion and argued that different cultures showed a diversity of preferences for attributes such as skin color, body hair, and body fat (Darwin as cited by [19]).

Despite cross-cultural and cross-gender differences in judgments of facial attractiveness, there is still a high level of agreement not only between individuals within a particular culture but also between individuals from different cultures or different genders [6, 20, 21]. In a comparison between judgments of 17 different social traits, as inferred from faces, facial attractiveness was found to have the highest interrater agreement and reliability level. The evidence for the existence of universal criteria for facial attractiveness raises two questions:

a. What is the functional role of facial attractiveness?

b. What are the facial diagnostic cues that constitute the criteria of attractiveness judgments?

As an answer to the first question, the evolutionary view suggests that attraction and repulsion to certain faces serve as an adaptive function [22]. Throughout evolutionary history, humans have developed preferences to specific phenotypical cues that lead them to choose mates who will provide the best chance of successful reproduction and survival of their own genes. The evolutionary approach has been based on the premise that an attractive face is a biological signal that provides valuable information about the quality of the signaler. Mate quality attributes may include characteristics such as health, fertility, intelligence, and potential for parental care. However, most research has focused on health ([23]; for review, see [24]).

To answer the second question, many studies have used facial image manipulations to test observers' responses. Some of these studies have suggested that there are several facial diagnostic cues that advertise the biological quality of an individual through the medium of the face. These phenotypical cues include: facial symmetry, averageness (i.e., faces that are not too far from the population mathematical mean of the geometric structure and texture) and sexual dimorphism (i.e., secondary sexual characteristics; see [25] for meta-analysis).

2. What makes a face attractive?

Despite a large body of research and findings, the question of what makes a face attractive is not easy to formalize. Among the reasons for that we can include the following nonexclusive list:

a. The complexity of the human face: the human face is a complex object and requires a high-dimensional data structure to represent and analyze it. Even if we want to represent information about a static non-expressive face, such data structure should include information about morphology (i.e., structure) and texture. Over recent decades, there has been a significant progress in the development of computational tools for the analysis and synthesis of faces [26–28].

b. The effect of the external physical environment: the viewing conditions may have a dramatic effect on the way we infer social traits from a face in general. Such physical conditions may include viewing distance, perspective, lighting conditions etc. It is not clear that the same facial determinants of social traits are identical under different viewing conditions.

c. Facial movements: the face is not a rigid object. A large set of groups of muscles (a.k.a. Action Units [29]) can and do change facial appearance. Many of these facial movements convey social signals such as emotional expressions. As a result, facial movements may affect the social inference from the face and even override the social impression of the default neutral and nonexpressive face. Facial movements add another level of complexity to the representation and analysis of faces; however, computational models for analysis and synthesis of facial movements are already in use [30].

Considering the above challenges, this chapter addresses the essence of facial beauty as a multifaceted question. To this end, we will approach the facial beauty signal as a part of a comprehensive communication system that comprises not only the signaling face as a transmitter but also the receiver (i.e., the observer) and the communication channel (i.e., of the external physical environment).

3. Attractiveness within the context of communication theory

All communications systems whether they are electronic, biological, or other comprise three fundamental elements:

a. The transmitter: the source that creates, modulates, and transmits the signals, for example a radio station.

b. The receiver: the agent that obtains the signal from the transmitter using a codebook. The codebook is an abstract list of rules that associates a meaning or reaction to specific messages.

c. The communication channel: the physical transmission medium or pathway that conveys the signals from the transmitter to the receiver; for example, a broadband fiber optic cable. A crucial point is that the communication channel modifies the transmitted signal. As a result, the signal received by the receiver is, in most cases, not identical to the original signal that was sent by the transmitter. Therefore, the physical properties of the channel determine the capability of the receiver to decode the transmitted signal [31]. In the case of social signaling from a face, the communication channel may have a large variety of characteristics: viewing distance, lighting conditions, the face or body's spatial orientation, partial occlusion, etc. The communication channel thus imposes constraints on the available information, changing the receiver's facial inference strategy. This means the question of what makes an attractive face is context dependent where a major factor that affects the attractiveness determinants is the communication channel.

3.1 The face as a signal transmitter

The face is a central communication tool in human social interaction. It transmits a large range of signals that convey social information to which the receiver associates meaning about the transmitter. This meaning whether it is reliable or not may include: gender, age, ethnicity, health condition, mood, intention, and competence. Some social impression signals (e.g., those indicating social traits such as dominance, trustworthiness, and attractiveness) are transmitted involuntarily by the default phenotypic morphology and complexion of the face [6]. However, other signals, such as facial expressions of emotion, can be voluntarily deployed strategically to negotiate social situations. Humans, as highly adaptive social beings and in a similar way to other social animals, can camouflage these involuntary morphology-based signals to boost chances of success within their ecological niche. In practice, humans deploy social-camouflage strategies by using dynamic facial signals to camouflage the involuntary social signals transmitted by static facial morphology [5]. In the latter study, using a computer graphics platform and a data-driven technique, facial action units (AUs, i.e., independent facial groups of muscles; [29]) were correlated with the impression of attractiveness to create a dynamic model of facial expression that elicits the impression of attractiveness. **Figure 1** depicts the facial movements that elicit the strongest and weakest intensities of attractiveness and lists the significant AUs that were combined to produce them. The color-coded heat maps show the movement magnitude of the 3-D vertices that make up each dynamic social signal. Using a similar approach, Gill et al. obtained dynamic models of facial movements that modulate the perception of two additional fundamental social traits: trustworthiness and dominance [5]. The latter study also examined the camouflaging capabilities of the modeled dynamic social gestures (i.e., whether the facial movements that are formalized in the models could override the involuntary default social signals transmitted by static facial morphology of the transmitter). The results revealed that attractiveness was the most difficult of these traits to camouflage. Humans are thus condemned to bear the social consequences of the inherited attractiveness of their faces. By contrast, social camouflage of dominance and trustworthiness is probably commonplace in everyday interactions. Casting directors are probably aware of this inequality. An attractive character will require an actor with attractive morphology; however, social camouflage can help an actor fake a dominant or trustworthy character.

3.2 Viewing distance as a communication channel

Social encounters may start at varying viewing distances. Based on the available information, individuals decide about their next action—whether to approach or avoid the other person. The face transmits a variety of social signals to receiving

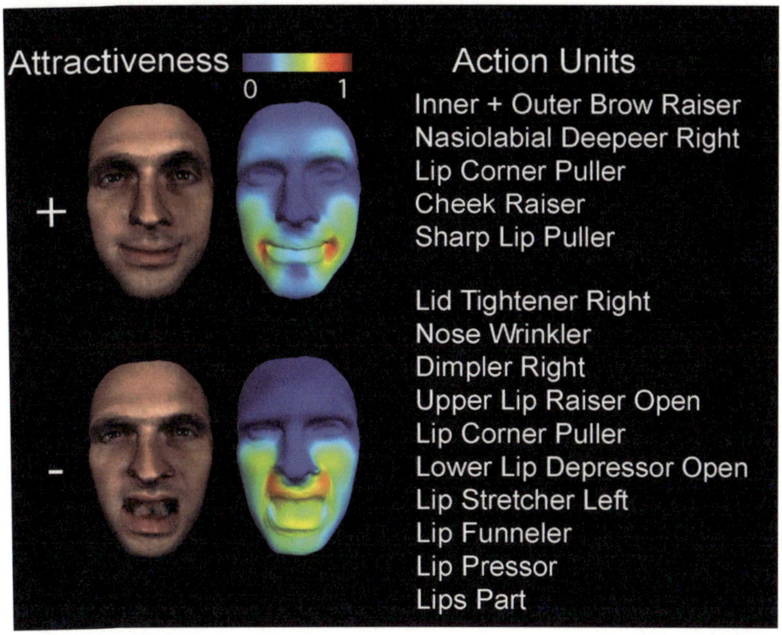

Figure 1.
Social camouflage. The two rows depict the signals of attractiveness with strong (+) and weak (−) intensities. The texture maps at the left illustrate the appearance of attractiveness on a common face. The color-coded heat maps indicate the location of dynamic face regions of the attractiveness signal; red indicates the highest magnitude of vertex movement. The column on the right lists the action units present in the majority of the observers' individual models [5].

observers across a wide range of viewing distances acting as a communication channel. Evaluating the social and reproductive capacity of others is paramount to negotiating the type of social interaction between individuals and, ultimately, to promoting the survival of the human species. However, social encounters begin at varying viewing distances, which can dramatically change the visual information and level of detail that is available for social judgments.

When a face moves closer to the receiver, its projection on the observing receiver's retina increases in size. As a result, the high spatial frequency information (HSF, representing fine details) projected initially on the retina progressively shifts toward lower spatial frequencies (LSFs, representing coarser scale and global information). Furthermore, new HSF details become progressively available on the retinal projection of the closer face for visual categorization in the receiver. When a face moves away from the receiver, it has the opposite effect: the retinal projection diminishes in size, the retina-based HSFs are no longer detected due to the finite resolution of the retina and the facial information initially represented in retina-based LSFs becomes retina-based HSFs. Combining different messages, transmitted by different spatial frequency bands, in one image is known as a *hybrid image* [32]. An illustrative example is shown in **Figure 2**, in which a hybrid image that combines the LSF of one face (a boy) with the HSF of another face (a girl). The available information of the image changes with viewing distance (or size) and as a result the face is perceived as a boy at a short distance (or in a large image size) and as a girl at a long distance (in a diminished size).

The critical impact of viewing distance, as a communication channel, raises the fundamental question of what specific facial signals communicate attractiveness and whether these signals change across viewing distance. Attractiveness diagnostic cues are found to covary with distance [33].

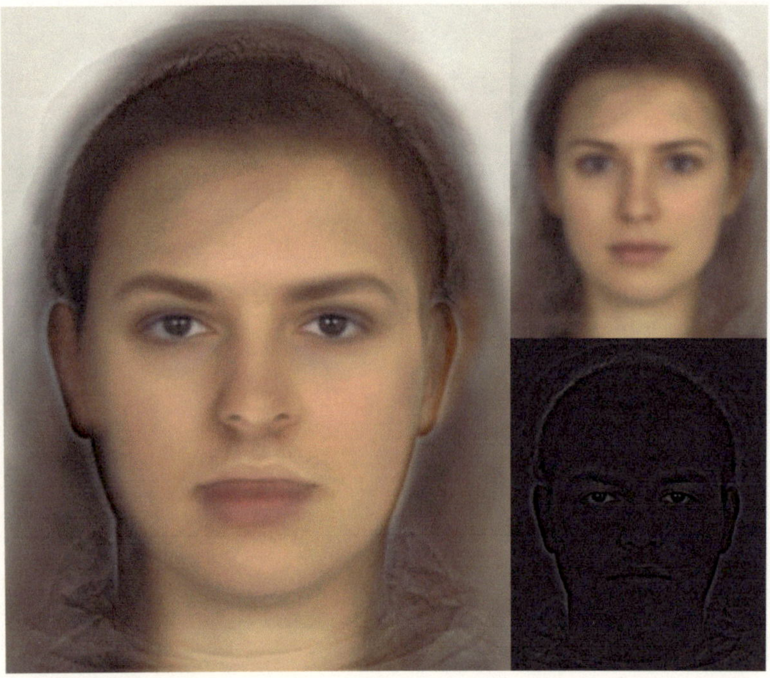

Figure 2.
Diagnostic cues vary with distance. Left panel: a hybrid image consisting of the low spatial frequencies (LSFs) of a girl and the high spatial frequency (HSF) of a boy. From a short viewing distance, the image is perceived as the face of a boy. From a long viewing distance, the image is perceived as the face of a girl. Right panel: the LSF of the girl's image (upper image) and the HSF of the boy's image (lower image).

Figure 3 illustrates how observers use different information from faces (both structure and texture) to assess attractiveness from proximal and distal signals. The diagnostic cues, both structural and textural, are color coded as follows: distal diagnostic cues are in red, proximal diagnostic cues are in green, and cues that are diagnostic in both distal and proximal distances are in yellow. Across all receiver-transmitter gender conditions (e.g., females observing males and females observing females) consistently, the hair has been found to be the prominent distal diagnostic cue. This is not surprising, considering the relatively large projection of the hair on the retina at long viewing distances compared to other facial attributes. Interestingly, among female transmitters, short or pulled-back hair is perceived as a distal signal of unattractiveness. The latter result suggests that masculine attributes in women (i.e., short hair) may distally signal characteristics associated with masculinity. Among male signals, trimmed hair is perceived as a distal signal of unattractiveness. In proximal viewing distance there are more available cues and the influence of the hair as a determinant cue decreases. Among both female and male transmitters, the eye region structure is found as a proximal diagnostic cue (though to different extent across receiver-transmitter's gender conditions and social traits). Moreover, among male transmitters, the glabellar frown lines were found as a proximal cue.

Another interesting question is whether viewing distance induces a natural hierarchy across different social traits in which humans infer some of these traits at longer viewing distances with greater sensitivity than other traits. The communication channel induces a natural hierarchy of decoding success, with attractiveness being the trait inferred from the greatest viewing distances. When comparing four basic social traits (aggressiveness, attractiveness, dominance, and trustworthiness), attractiveness was found to be inferred from the longest viewing distance tested (96 m, [33]).

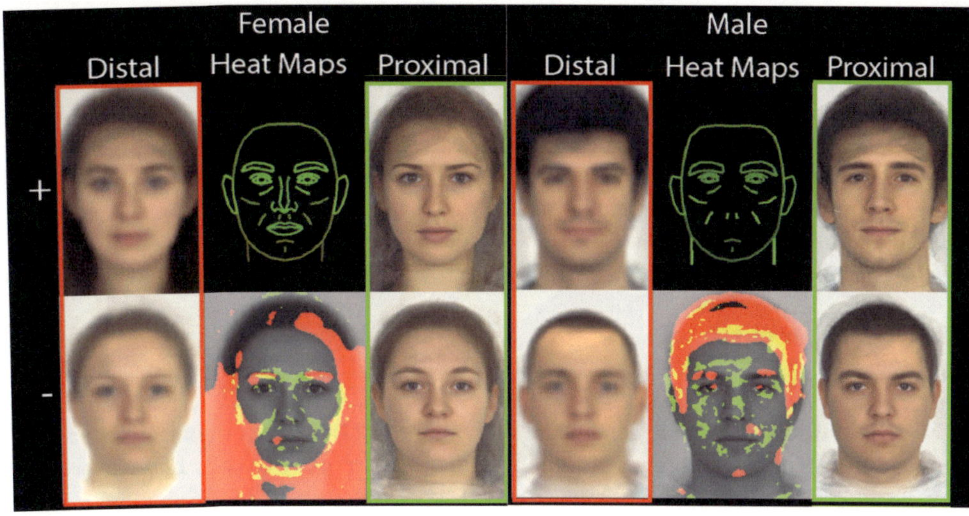

Figure 3.
Attractiveness predictions and diagnostic cues. The results are organized by sex of stimuli (columns) and based on judgment of opposite sex observers. For each sex of stimuli, the leftmost (distal—48 m) and rightmost (proximal—1.5 m) columns show the model predictions for the two polarities of trait denoted by + and – (attractive and unattractive correspondently). The middle column of each sex of stimuli category shows the diagnostic structural (i.e., shape of the facial features) and textural cues using color codes: green (exclusively distal cues), red (exclusively proximal cues), and yellow (both distal and proximal cues) [33].

4. Conclusions

The face is a communication tool that transmits a wide range of social signals. Inference of social traits from faces has evolved and diversified to serve, at least in part, as a rudimentary but instant communication tool to evaluate the benefits of cooperation or aversion during social interaction. Whether these signals are reliable or not, the facial signal decoding system in the receiver's brain has to deal with the major challenge of organizing the high-dimensional input from the retina and mapping it into a stable representation of a social category. Among the signals of some basic social traits, the signal of attractiveness is found to be outstanding in terms of interobserver agreement, reliability, and robustness to a variety of manipulations induced by the sender and the communication channel. Not only does the transmitter send complex social signals, in addition the communication channel induces large variability on the received signals. The limitations of the information that is caused by the camouflaging communication channel may challenge the receiver. An example for a potential scenario is that what is attractive from afar may be far from attractive at a short viewing distance and vice versa. The visual system is thus required to have a detailed set of diagnostic tools that varies with the availability of information and channel conditions. Interestingly, in a comparison among several social traits, attractiveness was found to be decoded from the longest distance, longer even the decoding distance of aggression.

The transmitter is not passive and by using specific facial movements they can camouflage the default neutral appearance of the face. Even if this is the case, the attractiveness of the face is found to be the most robust and hard to fake when compared with the other social traits that were studied [33].

Facial attractiveness is therefore a robust signal in social communication and the human brain seems to be adapted to detect it more effectively than any other social trait. There can be several possible reasons for the latter outcome. One possibility is that attractiveness signals provide more reliable information about the transmitter. For example, while attractiveness is a reliable signal of potential successful reproduction and survival of descendants, signals of trustworthiness may reflect actual levels of trustworthiness to a lesser extent (if at all). Another possibility is that with the limited computational and attentional capacities of the human brain, the higher sensitivity to attractiveness signals reflects the gain and loss priorities. Such risk management policy may give priority to approaching an attractive transmitter while ignoring hazard cues of aggression.

Acknowledgements

The author would like to thank Philippe G. Schyns, Benedict C. Jones, Oliver G. B. Garrod, and Rachael E. Jack from the University of Glasgow for their collaboration in some of the studies presented here. Special thanks to Stella E. Bell for her helpful suggestions during the write up of this chapter.

Author details

Daniel Gill
Department of Psychology, The University of Winchester, UK

*Address all correspondence to: daniel.gill@winchester.ac.uk

References

[1] Lennon JW, McCartney JP. I've Just Seen a Face, In: Help. London: EMI; 1965

[2] Buss DM. The Evolution of Desire: Strategies of Human Mating. 4th ed. New York: Basic Books; 2016. p. 448

[3] Langlois JH, Ritter JM, Roggman LA, Vaughn LS. Facial diversity and infant preferences for attractive faces. Developmental Psychology. 1991;**27**: 79-84. DOI: 10.1037/0012-1649.27.1.79

[4] Etcoff NL. Survival of the Prettiest: The Science of Beauty. 1st ed. New York: Anchor Books; 1999. p. 336

[5] Gill D, Garrod OGB, Jack RE, Schyns PG. Facial movements strategically camouflage involuntary social impressions of face morphology. Psychological Science. 2014;**25**(5): 1079-1086. DOI: 10.1177/0956797614522274

[6] Oosterhof NN, Todorov A. The functional basis of face evaluation. Proceedings of the National Academy of Sciences U.S.A. 2008;**105**:11087-11092. DOI: 10.1073/pnas.0805664105

[7] Stephen ID, Oldham FH, Perrett DI, Barton RA. Redness enhances perceived aggression, dominance and attractiveness in men's faces. Evolutionary Psychology. 2012;**10**(3):562-572. DOI: 10.1177/147470491201000312

[8] Ballew CC, Todorov A. Predicting political elections from rapid and unreflective face judgments. Proceedings of the National Academy of Sciences. 2007;**104**(46):17948-17953. DOI: 10.1073/pnas.0705435104

[9] Bar M, Neta M, Linz H. Very first impressions. Emotion. 2006;**6**(2):269-278. DOI: 10.1037/1528-3542.6.2.269

[10] Borkenau P, Brecke S, Möttig C, Paelecke M. Extraversion is accurately perceived after a 50-ms exposure to a face. Journal of Research in Personality. 2009;**43**(4):703-706. DOI: 10.1016/j.jrp.2009.03.007

[11] Todorov A, Pakrashi M, Oosterhof NN. Evaluating faces on trustworthiness after minimal time exposure. Social Cognition. 2009;**27**(6):813-833. DOI: 10.1521/soco.2009.27.6.813

[12] Willis J, Todorov A. First impressions: Making up your mind after a 100-ms exposure to a face. Psychological Science. 2006;**17**(7):592-598. DOI: 10.1111/j.1467-9280.2006.01750.x

[13] Flowe HD, Humphries JE. An examination of criminal face bias in a random sample of police lineups. Applied Cognitive Psychology. 2011;**25**:265-273. DOI: 10.1002/acp.1673

[14] Rezlescu C, Duchaine B, Olivola CY, Chater N. Unfakeable facial configurations affect strategic choices in trust games with or without information about past behavior. PLOS One. 2012;**7**(3):e34293. DOI: 10.1371/journal.pone.0034293

[15] Rule NO, Ambady N. The face of success: Inferences from chief executive officers' appearance predict company profits. Psychological Science. 2008;**19**:109-111. DOI: 10.1111/j.1467-9280.2008.02054.x

[16] Todorov A, Mandisodza AN, Goren A, Hall CA. Inferences of competence from faces predict election outcomes. Science. 2005;**308**:1623-1626. DOI: 10.1126/science.1110589

[17] Todorov A, Olivola CY, Dotsch R, Mende-Siedlecki P. Social attributions from faces: Determinants, consequences, accuracy, and functional

significance. Annual Review of Psychology. 2015;**66**:519-545. DOI: 10.1146/annurev-psych-113011-143831

[18] Rubenstein AJ, Langlois JH, Roggman LA. What makes a face at tractive and why: The role of averageness in defining facial beauty. In: Facial Attractiveness: Evolutionary, Cognitive and Social Perspective. Westport Connecticut: Ablex publishing; 2002

[19] Bruce V. Recognizing Faces. Hove and London: Lawrence Erlbaum; 1988. p. 154

[20] Cunningham MR, Roberts AR, Barbee AP, Druen PB. 'Their ideas of beauty are, on the whole, the same as ours': Consistency and variability in the cross-cultural perception of female attractiveness. Journal of Personality and Social Psychology. 1995;**68**:261-279. DOI: 10.1037/00223514.68.2.261

[21] Gill D. Women and men integrate facial information differently in appraising the beauty of a face. Evolution and Human Behavior. 2017;**38**(6):756-760. DOI: 10.1016/j.evolhumbehav.2017.07.001

[22] Zebrowitz-McArthur L, Baron RM. Toward an ecological approach to social perception. Psychological Review. 1983;**90**:215-238. DOI: 10.1037/0033-295X. 90.3.215

[23] Rhodes G, Yoshikawa S, Palermo R, Simmons WL, Peters M, Lee K, et al. Perceived health contributes to the attractiveness of facial symmetry, averageness, and sexual dimorphism. Perception. 2007;**36**:1244-1252. DOI: 10.1068/p5712

[24] Thornhill R, Gangestad SW. Facial attractiveness. Trends in Cognitive Science. 1999;**3**:452-460. DOI: 10.1016/S1364-6613(99)01403-5

[25] Rhodes G. The evolutionary psychology of facial beauty. Annual Review of Psychology. 2006;**57**: 199-226. DOI: 10.1146/ annurev. psych.57.102904.190208

[26] Blanz V, Vetter T. A morphable model for the synthesis of 3D faces. In: Proceedings of the 26th annual conference on Computer graphics and interactive techniques (SIGGRAPH '99); 8-13 August 1999; Los Angeles, New York: ACM Press; 1999. pp. 187-194. DOI: 10.1145/311535.311556

[27] Sirovich L, Kirby M. Low-dimensional procedure for the characterization of human faces. Journal of the Optical Society of America A. 1987;**4**(3):519-524. DOI: 10.1364/JOSAA.4.000519

[28] Turk M, Pentland A. Eigenfaces for recognition. Journal of Cognitive Neuroscience. 1991;**3**(1):71-86. DOI: 10.1162/jocn.1991.3.1.71

[29] Ekman P, Friesen WV. Facial Action Coding System: A Technique For The Measurement of Facial Movement. Palo Alto, CA: Consulting Psychologists Press; 1978. DOI: 10.1037/t27734-000

[30] Yu H, Garrod O, Schyns P. Perception-driven facial expression synthesis. Computers and Graphics. 2012;**36**(3):152-162. DOI: 10.1016/j.cag.2011.12.002

[31] Shannon CE, Weaver W. The Mathematical Theory of Communication. Urbana, Illinois: The University of Illinois Press; 1949. p. 117

[32] Schyns PG, Oliva A. Dr. Angry and Mr. Smile: When Categorization Flexibly Modifies the Perception of Faces in Rapid Visual Presentations. Cognition. 1999;**69**(3):243-265. DOI: 10.1016/s0010-0277(98)00069-9

[33] Gill D, Jones B. Facial diagnostic cues for social inferences covary with viewing distance [in preparation]

Chapter 10

Social Media and Its Effects on Beauty

Mavis Henriques and Debasis Patnaik

Abstract

Beauty is concerned with physical and mental health as both are intimately related. Short-term decisions to alter one's body structure irrespective of genetic, environmental, occupational and nutritional needs can leave medium- and long-term effects. This chapter analyzes the role of social media and its effects on the standards of beauty. The researchers have summarized the literature on how social media plays a role in affecting beauty trends, body image and self-esteem concerns. There is support that social media affects individuals negatively, in pushing them to engage in life threatening beauty trends due to social compliance and acceptance in society. The aim was to review social networking sites' impact on perception of standards of beauty and newer unrealistic trends gaining popularity that could alter opinions and also cause harm to individuals in the long run. This is an emerging area of research that is of high importance to the physical and mental health in the beauty, health and hospitality industry with the latter being manifested in depression, anxiety and fear of non-acceptability and being seen as a social gauche.

Keywords: social media, self-esteem, body modification, body dissatisfaction, beauty

1. Introduction

Social media refers to the use of websites and applications to create and share content or to participate in social networking [1]. Technological developments have given rise to various gadgets including smart-phones, tablets, and laptops to robots too. Living in a digitized era, communication has now become easier and faster with the emergence of various social applications available at the click of a button. While many may agree that social media has connected individuals globally, it has also been used to set standards of beauty for males, females as well as the third gender. This in turn has been known to affect the self-esteem of individuals with regards to body image, body modification and how they view themselves in society. In order to be accepted in society females have to battle body image issues from a very young age, where thin is considered to be the ideal body type [2].

This chapter focuses on the effects of social media on standards of beauty. We review the literature on the role of social media and how they affect the physical and psychological beauty of individuals in society.

2. Effects of social media on the aspects of body image

Today, Social Media is one of the most important factors contributing to the mental, emotional, physical and spiritual health of an individual. With the media constantly portraying ideal beauty and body image comparisons, the decisions of men and women's beauty choices are globally affected.

"Body image refers to a person's perception of their physical self and the thoughts and feelings, positive, negative or both, which result from that perception" [3]. Social media has had a major impact on the perceptual, affective, cognitive and behavioral aspects of body image [3] by encouraging lean body patterns and delivering anti-obesity messages [4]. Eating disorders determine a distorted relationship between the individual, their eating behavior and body shape [5]. Adolescence being a crucial age for positive and negative development of body image, the self-esteem and body dissatisfaction adolescents feel are known predictors of eating disorders [6]. Continuous pursuit for the perfect slender lean body may generate negative feelings which can result in a change in eating behavior, thereby increasing the chances of weight issues and eating disorders [4, 7]. Social media portrays women who are slim as being more beautiful and successful compared to overweight women [8]. Body image misperception and dissatisfaction with body weight highlight an association between body dissatisfaction and psychological wellbeing [9].

3. Self-esteem issues in response to social media effects

Body image concerns are common in women and men globally, but social media has now increased these concerns through advertising, videos and the use of social media. Milkie [10] conducted in-depth interviews on 60 white and minority girls to examine the effect of media on self-esteem. Results indicated that most girls felt that the images shown in media were unrealistic and not real. White girls felt that boys evaluated them on the basis of the images found on media platforms whereas the minority girls felt that the images portrayed on media did not meet the expectations of the reference group they oriented themselves with. The evolution from adolescence to adulthood has seen 12–16 teenage girls experience emotional changes in interpersonal and intrapersonal development as well as bodily changes such as sudden weight gain and transition from a young girl to a fully grown woman [11, 12]. In today's world, the self-presentation of beauty and perceptions of others plays an important role in developing identities in girls [13, 14]. New interactive platforms present in social media demonstrate how self-presentations and peer influences are interrelated with the standards of beauty [15]. Many women may imitate their ideal media personality due to the social, psychological and practical rewards associated with this ideal and the belief that their life would change for the better [16].

4. Social media: trends in behavioral outcomes

Social media comprises of social networking sites, image sharing sites, video hosting sites, community blogs, bookmarking sites and gaming sites. Fellow comparisons about self-image and appearances in teenagers have resulted due to social networking sites (SNSs) such as Instagram and Facebook [17]. Teenage girls engage in online self-presentation of posting selfies and sharing the outfit of the day pictures to differentiate themselves with their peers [18]. Media images of ideal beauty standards influence the content and sharing of pictures teenage girls' post [19]. Individuals are constantly seeking feedback on SNSs through likes, followers and comments to uphold a perfect and stable image of themselves [20]. Teenage girls are vulnerable to the upward comparison as it means that they need to improve their beauty standards, thereby leaving them dissatisfied with their physical bodies, having doubts about their self-worth and also driving them to self-harm behavior [20, 21].

Taking selfies and sharing them on popular social platforms such as Facebook, Instagram, Twitter and Snapchat has increased at an alarming rate during the recent years. A recent study compared selfie takers and non selfie takers and their perceptions of their selfies versus photographs clicked by others. Results indicated that selfie takers perceived themselves as more attractive and likable in their selfies as opposed to pictures taken by others leading to positive distortions of the self [22]. Biases in self-face recognition were seen in men and women in selecting the most attractive modified pictures of themselves [23]. Selfies are no harm per se. But obsession with physical features reveals a lack of holistic perception of self-generated sub-consciously, following an "outside" standard of beauty not defined by the "inner self" of the receiver.

Popular socialites Kim and Khloe Kardashian have been slammed with media reports of them using photoshop to edit Instagram selfies by making unrealistic alterations to look thinner and more toned. Emily Bryngelson, an associate designer struggling with an eating disorder, admitted to deleting pictures if they did not receive enough "likes." The time spent on Facebook photos was linked to self-objectification, weight dissatisfaction, thin idealization and pursuit of thinness [15].

5. Social media and unrealistic beauty standards

There are multiple factors that affect the beauty standards in the world today, which involve women and men and the third gender individuals trying new trends to be socially accepted. The purchasing decisions of millennials are influenced majorly by social media [24]. 72% of millennials procure beauty products based on Instagram posts and other social networks [25]. Makeup consists of the application of cosmetic products to beautify or change the way one looks either artistically or to conceal flaws. Jang-Soon and Hye-Jin [26] investigated 240 teenage males' preferences for makeup use. Results indicated that their appearance was one of the major reasons for their social success. The male respondents who were young, unmarried and city residents had an overall positive perception about cosmetics [26]. With bloggers constantly advertising on social media, cosmetic products, have gained popularity.

The images on social media sites are idealized and unreal, due to digital alteration thereby setting high expectations from individuals in society. Imperfections are removed by airbrushing and using other digitized apps to whiten teeth, slim waists and reduce sizes in order to be accepted as beauty ideals [27]. These techniques may further lead to negative consequences of increased body dissatisfaction, body modification and low self-esteem issues. Unrealistic images of feminity, beauty, success and body shape promoted through social media images are associated with development of eating disorders and body dissatisfaction disorders [28, 29].

Filters and beauty apps represent another area in which social media has a major influence. Beauty apps encourage women to see and surveil themselves within a "pedagogy of defect" [30]. They include filters and modification apps, surgery try out apps, and esthetic benchmarking apps which help individuals visualize how they will look after certain changes such as teeth whitening, eye bag removal and also whether the individual looks old or young [31].

"Body modification refers to the deliberate or permanent altering of an individual's human anatomy or appearance" [32]. They involve two aspects: the processes that modify form or contours of the body such as metabolic manipulation (weight lifting, extreme dieting, use of drugs/steroids, hormones), cosmetic surgeries and procedures (liposuction, face-lifts, rhinoplasty, botox, eye lash extensions), genital surgery and sex reassignment surgery, restriction or compression (waist training, foot binding), abrasion (teeth filing, scourging, flagellation), elongation (neck, lips, earlobes), partial or full removal of body parts (breasts, penis, ribs, nose etc.), implantation of foreign objects (silicone implants, decorative items under the skin), and prosthetics (false limbs, finger nails, lenses) and processes that mark the surface of the body such as tattooing, piercing, tanning/bleaching, scarification, branding and hair removal [32].

Young women and teenage girls following fitness boards on Pinterest were likely to have intensions to engage in extreme crash dieting or extreme exercising as a result of social comparison leading them to feelings of inadequacy and body dissatisfaction [33, 34].

6. Body modification trends in society

From professional athletes to celebrities, contouring, tattooing and body piercings has gained popularity in today's society [35]. Individuals who get tattooed refer to it as a piece of art and piercing as fashion accessories, for the purpose of embellishment or as a self-healing effect after having being abused [36–39]. A survey conducted at an American University found body piercings in 42% of men and 60% of women with piercings involving tongue, lips, nose, navel, genitals, nipple and eyebrow besides the earlobe piercing. Bacterial infections, bleeding and local trauma were common complications faced. "Tattoos were present in 22% of male students and 26% of female students" [40].

Another reason why individuals engage in body modifications is to maintain self-identities and be distinctive from others [41, 42]. Physical endurance, lust for pain, spirituality and cultural traditions, addictions, resistance, sexual motives, group commitments are reasons why individuals adopt modification procedures [43]. Social Media has had a tremendous effect on how individuals perceive and endure painful tattooing and body modifications after viewing popular Instagram and Pinterest handles. Brief exposure to body modifications on popular social networking sites has seen an increase in tattoo searches and body modification procedures in young as well as older individuals, proving the impact of Social Media on Beauty trends in society.

7. Adopting to ever changing social media trends: Is it right?

Social media has a robust influence on the beauty, health and hospitality industry with women and men engaging in weight loss and diets to avoid gaining fat identities that impacted their wellbeing in the long run [44–46]. Women and men have turned to waxing, shaving and removal of unwanted facial and body hair in order to meet the beauty standards of societal acceptance [47, 48]. Women who did not engage in hair removal were negatively evaluated as being dirty or gross [49, 50]. In 2010, a concept of living dolls emerged online with women practicing the art of appearing "doll like." These women would engage in usage of wide rimmed contact lenses, hair extensions, corsets, photo editing and surgery including, eye widening, breast implants, liposuction and rib removal to enhance their beauty [31]. Even though women knew the risks in false eyelashes and acrylic nails, they still reported to be continuing to engage in it to feel socially accepted [31].

8. Using social media for body positivity

Social networking sites such as Facebook, Instagram, Twitter and other networking sites have the potential to influence positive beliefs and attitudes in individuals [51]. The online platform has given many individuals a feeling of a "sense of belonging." Men and women are obsessed with images on social media portals and often search for esthetic body types which are not similar to their own body. Instagram and Facebook often have stories of individuals who have fought hard to change beauty standards through sheer dedication and hard work, be it exercising, eating healthy or building self-esteem and body acceptance through support groups and communities found online. Blocking body shamers can help reinforce confidence in men and women globally. Promoting videos on life struggles with weight motivate others to believe in never giving up and taking charge of their lives. Today social media includes individuals of different race, gender, ethnicity and sexual orientation, thereby focusing more on breaking stereotypes and building communities to support each other.

9. Conclusion

With technology advancing by the minute and newer apps surfacing online, social media has an immediate effect on beauty. Due to the ever changing body images depicted online, individuals are turning to social media handles for acceptance and support. The selfie culture has brought about a positive and negative change in how individuals perceive themselves. While most research today focuses on the negative impacts of social media on beauty, more interest should be laid on body positivity and using social media as a medium for self-acceptance whether beautified or not.

Author details

Mavis Henriques[1*] and Debasis Patnaik[2]

1 Birla Institute of Technology and Science, Goa Campus, Zuarinagar, Goa, India

2 Department of Economics, Birla Institute of Technology and Science, Goa Campus, Zuarinagar, Goa, India

*Address all correspondence to: mavis_henriques@hotmail.com

References

[1] Social Media [Internet]. 2020. Available from: https://www.lexico.com/definition/social_media

[2] Grabe S, Ward ML, Hyde JS. The role of the media in body image concerns among women: A meta-analysis of experimental and correlational studies. Psychological Bulletin. 2008;**134**(3):460-476. DOI: 10.1037/0033-2909.134.3.460

[3] Fardouly J. Social media and body image. NEDC e-Bulletin. 2020;**46**:1-4

[4] Cecon RS, Franceschini SCC, Peluzio MCG, Hermsdorff HHM, Priore SE. Overweight and body image perception in adolescents with triage of eating disorders. The Scientific World Journal. 2017;**2017**:8257329. DOI: 10.1155/2017/8257329

[5] Cavalcanti AMTS, de Arruda IKG, de Lima EACM. Characterization of eating behavior disorders in school-aged children and adolescents: A population-based study. International Journal of Adolescent Medicine and Health. 2016;**29**(3):1-8. DOI: 10.1515/ijamh-2015-0087

[6] Voelkar DK, Reel JJ, Greenleaf C. Weight status and body image perceptions in adolescents: Current perspectives. Adolescent Health, Medicine and Therapeutics. 2015;**6**:149-158. DOI: 10.2147/AHMT.S68344

[7] Lofrano-Prado MC, Prado WL, De Piano A, Damaso AR. Obesidade e transtornos alimentares: A coexistência de comportamentos alimentares extremos em adolescentes. ConScientiae Saúde. 2011;**10**(3):579-585. DOI: 10.1016/j.eatbeh.2010.09.008

[8] Toriola AL, Dolan BM, Evans C, Adetimole O. Weight satisfaction of Nigerian women in Nigeria and Britain: Intergenerational and cross-cultural influences. European Eating Disorders Review. 1996;**4**(2):84-94. DOI: 10.1002/(sici)1099-0968(199606)4:2<84::aid-erv148>3.0.co;2-w

[9] Moehlecke M, Blume CA, Cureau FV, Kieling C, Schaan B. Self-perceived body image, dissatisfaction with body weight and nutritional status of Brazilian adolescents: A nationwide study. Journal of Pediatria. 2018;**96**(1)1-8. DOI: 10.1016/j.jped.2018.07.006

[10] Milkie MA. Social comparisons, reflected appraisals, and mass media: The impact of pervasive beauty images on black and white girls' self-concepts. Social Psychology Quarterly. 1999;**62**(2):190. DOI: 10.2307/2695857

[11] Labre MP, Walsh-Childers K. Friendly advice? Beauty messages in websites of teen magazines. Mass Communication and Society. 2003;**6**(4):379-396. DOI: 10.1207/S15327825MCS0604_3

[12] Vandenbosch L, Eggermont S. Understanding sexual objectification: A comprehensive approach toward media exposure and girls' internalization of beauty ideals, self-objectification, and body surveillance. Journal of Communication. 2012;**62**(5):869-887. DOI: 10.1111/j.1460-2466.2012.01667.x

[13] Caspi A. The child is father of the man: Personality continuities from childhood to adulthood. Journal of Personality and Social Psychology. 2000;**78**(1):158-172. DOI: 10.1037//0022-3514.78.1.158

[14] Martin MC, Kennedy PF. Advertising and social comparison: Consequences for female preadolescents and adolescents. Psychology and Marketing. 1993;**10**(6):513-530. DOI: 10.1002/mar.4220100605

[15] Meier EP, Gray J. Facebook photo activity associated with body image disturbance in adolescent girls. Cyberpsychology, Behavior and Social Networking. 2014;**17**:199-206. DOI: 10.1089/cyber.2013.0305

[16] Engeln-Maddox R. Buying a beauty standard or dreaming of a new life? Expectations associated with media ideals. Psychology of Women Quarterly. 2006;**30**(3):258-266. DOI: 10.1111%2Fj.1471-6402.2006.00294.x

[17] Mascheroni G, Vincent J, Jimenez E. Girls are addicted to likes so they post semi-naked selfies: Peer mediation, normativity and the construction of identity online. Cyberpsychology: Journal of Psychosocial Research on Cyberspace. 2015;**9**(1):5. DOI: 10.5817/CP2015-1-5

[18] Kaplan AM, Haenlein M. Users of the world, unite! The challenges and opportunities of social media. Business Horizons. 2010;**53**:59-68. DOI: 10.1016/j.bushor.2009.09.003

[19] Boyd D. It's Complicated: The Social Lives of Networked Teens. New Haven, CT: Yale University Press; 2014

[20] Corcoran K, Crusius J, Mussweiler T. Social comparison: Motives, standards, and mechanisms. In: Chadee D, editor. Theories in Social Psychology. Oxford: Wiley-Blackwell; 2011. pp. 119-139. DOI: 10.1037/a0023922

[21] Chua THH, Chang L. Follow me and like my beautiful selfies: Singapore teenage girls' engagement in self-presentation and peer comparison on social media. Computers in Human Behavior. 2016;**55**:190-197. DOI: 10.1016/j.chb.2015.09.011

[22] Epley N, Whitchurch E. Mirror, mirror on the wall: Enhancement in self-recognition. Personality and Social Psychology Bulletin. 2008;**34**:1159-1170. DOI: 10.1177/0146167208318601

[23] Wen W, Kawabata H. Why am I not photogenic? Differences in face memory for the self and others. Perception. 2014;**5**:176-187. DOI: 10.1068/i0634

[24] Likhareva E, Kulpin S. Social media influence on consumption in the beauty industry: Modern studies. UDC. 2018;**304**:1-6

[25] Social media influence on sales [Internet]. 2017. Available from: https://www.forbes.com/sites/andrewarnold/2017/12/22/4-ways-social-media-influencesmillennials-purchasing-decisions/#3cdd48ea539f

[26] Jang-Soon P, Hye-Jin K. Perception about makeup influence on Man's makeup and their success. Journal of the Korea Convergence Society. 2017;**8**(4):231-237

[27] Farid H. Seeing is not believing. IEEE Spectrum Archive. 2009;**46**(8):44-48. DOI: 10.1109/MSPEC.2009.5186556

[28] Hesse-Biber S, Leavy P, Quinn CE, Zoino J. The mass marketing of disordered eating and eating disorders: The social psychology of women, thinness, and culture. Women's Studies International Forum. 2006;**29**:208-224. DOI: 10.1016/j.wsif.2006.03.007

[29] Grogan S. Body Image: Understanding Body Dissatisfaction in Men, Women and Children. 2nd ed. East Sussex: Routledge; 2008

[30] Bordo S. Twilight Zones: The Hidden Life of Cultural Images from Plato to O.J. Berkeley: University of California Press; 1997

[31] Elias AS, Gill R, Scharff C. Aesthetic Labour: Rethinking Beauty Politics in Neoliberalism. London: MacMillan Publishers; 2017

[32] Hayes K. Body modifications. Encyclopedia of sex and gender:

Culture society history. Social Sciences. 2020;**1**:1-6

[33] Lewallen J, Behm-Morawitz E. Pinterest or thinterest?: Social comparison and body image on social media. Social Media and Society. 2016;**2**:1-9. DOI: 10.1177%2F2056305116640559

[34] Alperstein N. Social comparison of idealized female images and the curation of self on Pinterest. The Journal of Social Media in Society. 2015;**4**:5-27. Available from: https://www.thejsms.org/index.php/TSMRI/article/view/82

[35] Heidelbaug JJ. Clinical Men's Health: Evidence in Practice. Elsevier Health Sciences. 2008;**17**:323-338

[36] Wohlrab S, Stahl J, Kappeler PM. Modifying the body: Motivations for getting tattooed and pierced. Body Image. 2007;**4**(1):87-95. DOI: 10.1016/j.bodyim.2006.12.001

[37] Atkinson M. Pretty in ink: Conformity, resistance, and negotiation in women's tattooing. Sex Roles. 2002;**47**:219-235. DOI: 10.1023/A:1021330609522

[38] Stirn A. Body piercing: Medical consequences and psychological motivations. The Lancet. 2003;**361**:1205-1215. DOI: 10.1016/S0140-6736(03)12955-8

[39] Stirn A. Motivations of tattooed and pierced for their body modifications. Zeitschrift fu r Klinische Psychologie, Psychiatrie und Psychotherapie. 2004;**51**:43-58

[40] Mayers LB, Judelson DA, Moriarty BW, Rundell KW. Prevalence of body art (body piercing and tattooing) in university undergraduates and incidence of medical complications. Mayo Clinic Proceedings. 2002;**7**(1):29-34. DOI: 10.4065/77.1.29

[41] Millner VS, Eichold BH. Body piercing and tattooing perspectives. Clinical Nursing Research. 2001;**10**:424-441. DOI: 10.1177/C10N4R7

[42] Stirn A. Designing the own body – Narcissistic aspects of tattoos and piercings. Psychotherapie im Dialog. 2004;**3**:256-260

[43] Stirn A, Peregrinova L, Oddo S, Philipp S, Hinz A. Motivations for body piercings and tattoos- the role of sexual abuse and the frequency of body modifications. Psychiatry Research. 2011;**190**:359-363. DOI: 10.1016/j.psychres.2011.06.001

[44] Dalley SE, Buunk AP. 'Thinspiration' vs. 'fear of fat': Using prototypes to predict frequent weight-loss dieting in females. Appetite. 2009;**52**(1):217-221

[45] Vartanian LR. Disgust and perceived control in attitudes toward obese people. International Journal of Obesity. 2010;**34**(8):1302-1307. DOI: 10.1038/ijo.2010.45

[46] Wray S, Deery R. The medicalization of body size and women's healthcare. Health Care for Women International. 2008;**29**(3):227-243. DOI: 10.1080/07399330701738291

[47] Tiggemann M, Lewis C. Attitudes toward women's body hair: Relationship with disgust sensitivity. Psychology of Women Quarterly. 2004;**28**(4):381-387. DOI: 10.1111%2Fj.1471-6402.2004.00155.x

[48] Toerien M, Wilkinson S, Choi PYL. Body hair removal: The 'mundane' production of normative femininity. Sex Roles. 2005;*52*(5-6):399-406

[49] Toerien M, Wilkinson S. Gender and body hair: Constructing the feminine woman. Women's Studies International Forum. 2003;**26**(4):333-344

[50] Toerien M, Wilkinson S. Exploring the depilation norm: A qualitative questionnaire study of women's body hair removal. Qualitative Research in Psychology. 2004;**1**(1):69-92. DOI: 10.1191/1478088704qp006oa

[51] Perloff RM. Social media effects on young women's body image concerns: Theoretical perspectives and an agenda for research. Sex Roles. 2014;**71**:363-377. DOI: 10.1007/s11199-014-0384-6

Chapter 11

Blossoming for Whom? Social Approval and Body Image

Wenting Mu and Fan Wu

Abstract

Body image is a multidimensional construct that reflects the way we perceive and feel about our physical appearance. This inside view of our body heavily influences our self-esteem, mental health, and overall well-being. Under the influence of mass media, peers and family, individuals, especially women, may feel pressured to conform to the societal standards of beauty, engage in upward social comparison, and consequently experience negative body image. While our sociocultural surroundings plays a role in the internalization process, other intrapersonal factors, such as appearance-based rejection sensitivity and lack of self-concept clarity, may heighten the risk for some individuals. Body image disturbances can be manifested in forms of avoidance behaviors, monitoring, eating restraints, and body modification. In order to promote body acceptance, we ought to gain insights into the formation of our body image and challenge the commonly held belief on who defines beauty.

Keywords: body image, societal standards, media message, social approval

1. Introduction

Our relationship with our body has always been complicated. Although our body is an integral part of our life experience, many of us do not grow to like or appreciate our body as the way it is. Through the process of socialization, we develop ideas of what is preferred in our society and how we should look at our body. This belief system shapes our perception of and attitudes towards our body. While the size and shape of our body are an objective measure, our body image is not always as stable or realistic. It is not uncommon for one to hold a distorted view of his or her body, which could lead to body dissatisfaction and compromise one's psychological well-being.

Psychologists have devoted an entire body of research on this particular topic and investigated how our perception of our body influences thoughts, beliefs, feelings, and behaviors. Some researchers looked into the origins of body image, others studied the influence of negative body image on our mental health. They raised many interesting questions that inspire us to reflect upon. For example, what role do our families and friends play in the shaping of our body image? What makes some individuals more vulnerable to social comparison than others? In this chapter, we will delve into some of these questions and reveal how our societal standards explicitly and implicitly influence our body image.

Before we get started, it is worth mentioning that body image is not a subject that only pertains to girls or women. As many of our male readers may attest, men can be equally troubled by their body images as well. Moreover, body image is not only about size. It would be unfair to assume that one is only concerned with his or her weight or size, given the rich diversity of human experiences. Literature has examined a great variety of body-related topics, such as physical diseases and injuries. For the purpose of this chapter, we will mainly look at the section of literature that relates to shapes and sizes. But it is not to say that other aspects of body image are not as equally important.

2. Body image and beauty

Body image is commonly understood as how one looks in the mirror. By staring at your body in the mirror, you will notice some physical characteristics including shape, size, height, skin tone, and so forth. However, do you think these characteristics truly reflect your body? Does your body seem bigger or smaller than you expect? I think you would agree with me that what we see is often not how we feel. There is a great amount of discrepancy between what our body actually looks like and what we perceive our body to be. Furthermore, how do you like the way you look? Do you have an opinion about your body every time you check yourself out in the mirror?

Psychologists coined the term body image to refer to one's perceptions and attitudes towards his/her physical characteristics [1]. Body image is a multidimensional concept that subsumes cognitive, affective, behavioral, and perceptual facets [2, 3]. For instance, perceiving your body to be a certain way can give rise to various emotions, lead you to have positive or negative thoughts, and result in behaviors in an attempt to modify your body.

More specifically, psychologists have been interested in studying the development of body image and its influence on other aspects of people's life. Plato once said, "The body, in which we are imprisoned like an oyster in its shell" [4]. One can easily imagine that our feelings towards and opinions of our bodies can fundamentally influence our day-to-day life experiences. Therefore, in order to understand one's idea of beauty, we must inevitably take a close look at his/her body image.

The first question that researchers began to investigate was how accurate people's body perception was. Researchers asked individuals to estimate their body sizes and compared the estimations against their actual measurements. The results show that some people have inaccurate estimations as they perceived their bodies to be either bigger or smaller than their actual sizes [5, 6]. Furthermore, people who tend to overestimate their sizes are more likely to develop eating disorders [7]. Among anorexic patients, researchers examined whether their body distortions stem from inaccurate visual inputs or distorted views of their bodies. They employed assessments such as digital photography techniques and figure drawing scales. It turns out that there is not much difference in the patients' visual sensitivity (i.e. heightened ability in processing visual information) but disturbances in how they interpret the images of their bodies [8, 9]. Hence, their biased attitudes towards weight and size caused them to have a distorted body image.

This discovery naturally led to the next question about the attitudinal component of body image. In other words, what attitudes do people have in relation to their physical appearance? Researchers examined this question by first asking people if they were satisfied with their body image. The results show that around 61 to 93% of people were not satisfied with either their overall appearance or specific body areas [10, 11]. More specifically, around 50% of preadolescent girls and 30% of preadolescent boys reported dissatisfaction with their body [12–14]. In adults, approximately 60% of women and 40% of men see their body negatively, and these rates remain stable across the lifespan [15, 16]. In addition to being dissatisfied, people also reported experiencing emotional distress (including shame, anxiety, or discomfort) with regards to their body [1]. They could experience such distress at specific moments or as part of their general life experience.

Body image distortion and dissatisfaction can have serious consequences. People with negative body image are at risk of having low self-esteem, depression, social anxiety, impaired sexual functioning, and reduced quality of life [1, 17, 18]. They may engage in risky health behaviors including unhealthy eating, physical inactivity, unsafe sex, smoking, and so forth [19–22]. Negative body image can also contribute to the development and maintenance of body dysmorphic disorder and eating disorders [23, 24].

Eating disorders are serious and sometimes life-threatening illnesses. They often involve serious medical complications that can cause permanent damage or death. People with eating disorders also have an increased risk of dying by suicide. According to the National Institute of Mental Health, the accumulated lifetime prevalence of eating disorders (including anorexia nervosa, bulimia nervosa, and binge eating disorder) was around 4% among adults and 2.7% among adolescents aged 13 to 18 years [25]. Overall, approximately 30 million Americans have struggled with an eating disorder over their lifetime. Moreover, probably twice the number of people or more are also struggling with eating disturbances even if their conditions do not yet meet the criterion of a clinical diagnosis [26]. This is why body image distortion and dissatisfaction are implicated in a range of public health concerns such as eating disorders [27].

For people with negative body image, their passage to beauty and self-appreciation is blocked, because having a healthy and positive body image sits at the core of beauty. Merriam-Webster defines beauty as "a quality or aggregate of qualities in a person or thing that gives pleasure to the senses or pleasurably exalts the mind or spirit" [28]. It is one thing to please the senses of others, but it is another to please oneself. To appreciate the beauty within him/herself, a person must be able to view his/her body positively in the first place. However, it would difficult if a person views him/herself unfavorably and struggles with unpleasant emotions and feelings towards his/her own body. Therefore, in order to reinstate the sense of beauty and self-appreciation within individuals, we must first understand the concept of body image.

3. Societal standards of body image

At this point, you might wonder what causes people to have distorted and disapproving views of their body. Thomas F. Cash, a leading expert in the field of body image, proposed that there are two views of human appearance [29]. One is the "outside view" as how our physical appearance influences our interpersonal experiences. For example, physical attractiveness plays a role in an array of contexts such as friendship, romantic relationships, and job opportunities. The other is the "inside view" which is a person's subjective experiences of his/her appearance. The inside view was later defined as "body image" [29]. When we talk about body image, we are referring to a person's own perceptions, attitudes, emotions, and beliefs regarding his/her appearance. However, the inside view is built upon and largely influenced by the outside view.

Societal standards regarding body image have a prominent influence on an individual's self-image. As a society, we hold standards for a large variety of qualities. Among these dimensions of self, one's physical body is the most prominent [30]. The values and beliefs about physical attractiveness are referred to as societal standards of beauty. The concept of beauty is ever-changing, as it has constantly evolved over time and varied across cultures. The standards of different societies stem from two main sources of influence: biology and culture [31]. In developed countries, as the issue of survival becomes less of a concern, people's preferences for ideal body shape have shifted from a sultry and voluptuous ideal to a thin and slender ideal. Nowadays, the modern ideal body shape has increasingly become both thin and very fit [32]. These ideals then circulate to other parts of the world due to globalization and have been infused into the standards of other countries and cultures. A cross-national comparison study found that the pressure to conform to Western ideals significantly predicts ideal body stereotype internalization for female participants from America, Poland, and the Czech Republic [32].

How are people influenced by these societal standards? Psychologists proposed a persuasive model - the Tripartite Influence Model [33]. It describes how social influence from media, family, and peers can predict body image and eating disturbances. The model also suggests that individual factors, such as internalization of the ideal body shape (regardless of how the ideal body shape is defined in a given society) and a chronic tendency towards social comparison, could mediate these social influences.

Societal standards are pervasively communicated through media messages. Traditional forms of media, such as TV commercials and magazines, have been advocating and promoting the desirability of an unrealistically thin ideal [34, 35]. For example, a study about print media found that adolescent girls would endorse their ideal as the models in fashion magazines specifically targeting teenage girls [36]. New social media, such as Instagram, Pinterest, or Tumblr, are image based. Seeing the images of thin and athletic peers provides a convenient target of upward social comparison for female viewers and motivates them to achieve a similar body shape. Many researchers in the field of eating disorders have criticized the media's role in the formation of eating and body image disturbances (e.g., [34–36]).

The media equate an ideal body shape not only with attractiveness but also with success. American culture emphasizes that physical attractiveness helps to achieve success in every area of life [32]. It is believed that thinness is crucial for success and happiness and people with the ideal body shape are likely to have a high social status. On the contrary, overweight or obese people are under pervasive appearance-based social discrimination and are often associated with negative qualities such as unattractive, lazy, immoral and dishonest [37].

Through this socialization process, women are disproportionately influenced by the social standards of beauty. Studies have unanimously observed greater body dissatisfaction among women than men [8, 9, 15]. Psychologists propose that the gender differences in body image originate from a sexual dimorphism through the general developmental process and the subsequent divergent psychosocial experiences of both genders [8, 31]. From an evolutionary perspective, beauty and attractiveness are not merely a cultural concept, but rather an important factor in determining one's odds of survival and reproductive success. Through the process of defining prominent features of attractiveness for both genders, different standards have emerged. Traditional gender roles associate femininity with beauty and the desire for an attractive appearance, while masculinity is associated with force and control [8]. This focus on esthetic qualities of the body creates a low level of body esteem and dissatisfaction among women [31]. Moreover, mass media portray beauty as a woman's primary objective. They also normalize the pressure that women experience with body image as if it is normal and acceptable for a woman to be ashamed and anxious about her body and appearance [38]. However, this is by no means to say that men are not affected by societal standards of beauty. In fact, a growing trend of body dissatisfaction has been observed among men [39].

While some of them are affected by the thin ideal, others are actively pursuing a masculine body ideal with high muscle mass. This tendency could posit men at similar risk of developing body image disturbance and eating disorders.

4. Internalization of societal standards

Not all women are equally influenced by societal standards of beauty. Some individuals may not be affected by media messages at all, while others are greatly affected and tend to modify their behaviors in a dysfunctional way to model media-promoted images. Researchers speculated that some interpersonal and intrapersonal factors play a role in influencing an individual's response to societal standards.

One specific individual difference variable, internalization of societal pressures regarding standards of attractiveness, appears to moderate or even mediate the media's effects on women's body dissatisfaction and eating dysfunction. Internalization is a process in which individuals assimilate the idea that possessing the ideal body shape is associated with being happy and successful and begin to hold themselves up against the societal standards [40]. Researchers identified two trends of internalization. One is thin-internalization, in which an individual wants to be thin or skinny; the other is athlete-internalization referring to the desire to possess a lean but muscular body [33]. While these two trends differ in the societal standards that one subscribes to, individuals who experience greater pressure to conform to either societal standard are more likely to internalize the ideal body stereotype [39]. Living in a culture that puts such a strong emphasis on the media and uses it to convey beauty standards increases the perceived pressure to conform for some individuals. This may increase their propensity to internalization and create a strong desire for them to achieve ideal beauty standards. Thus, they are at a greater risk of experiencing body image disturbances.

Empirical studies substantiate extensive evidence of internalization. For example, middle-school-aged girls who perceive higher peer influence and more television influence on the importance of attractiveness reported greater dissatisfaction with their body image and having pathological beliefs about eating [14]. College women who saw photographs of thin models from fashion magazines reported significantly higher levels of private body self-consciousness and anxiety than women in the control group [41]. Women with eating disorders demonstrated significant increases in overestimating their body size following exposure to photographs of models from popular fashion magazines. However, no such effect was found among women without eating disorders [41]. People with bulimic symptoms, regardless of the severity of their symptoms, experienced lower levels of self-esteem and weight satisfaction after seeing photographs of thinner models compared to after seeing photographs of larger models [42].

Why are some individuals more likely to internalize societal standards than others? Psychologists suggest that "People differ considerably in how they want to be seen, but they share in common an active pursuit of those desired self-images." [30]. Femininity and attractiveness, among many other qualities, are important to women's self-image. One study found that, in order to present themselves as feminine, female participants consumed less food when getting acquainted with a desirable male companion than with others [43]. Compared to the other qualities, societal standards of attractiveness is a highly accessible external source that can be used to define the self. Therefore, for women who lack a clear definition of their identity, internalizing the thin ideal and comparing their appearance to other women may serve as a means to gain self-knowledge [44].

Unfortunately, defining their self-concept in these ways can have negative implications for women's body image. A sociocultural model stressed that the current societal standards for thinness, as well as other standards of beauty for women are impossible to achieve for the average woman [31, 34]. It is not difficult to notice discrepancies between the actual self and the ideal self through upward social comparison. For example, when flipping through fashion magazines, one may make comparisons between her body shape and the models' and inevitably notice the differences in their sizes. Strauman and colleagues found that actual-ideal discrepancies among female undergraduates were correlated with dissatisfaction with weight and appearance-related beliefs about self [45].

Another source of self-knowledge comes from one's interaction with others. Interpersonal contexts provide opportunities for one to gain insights into his/her self-knowledge [30]. In particular, they allow people to choose comparison targets and interaction partners in ways that maximize benefits to the self. People tend to choose interaction partners who see them as they see themselves due to the desire for self-verification [46]. They create environments that confirm their self-views, primarily by choosing appropriate interaction partners, and they interpret and remember their interactions as confirming their self-views. People choose and are highly committed to interaction partners who confirm their self-views, even if those self-views are negative [46]. Therefore, their negative self-views can be verified, maintained, or even reinforced through their interactions.

5. Social approval and fear of appearance-related rejection

Messages about societal standards of attractiveness do not emanate just from media sources. Unfortunately, family, peers, coaches, teachers, and others help reinforce this socialization of women [37, 47, 48]. Their perception plays a critical role in influencing our self-image. Many interpersonal influences have been identified to contribute to the development and maintenance of shape- and weight-related disorders [49]. The factors include, but are not limited to, teasing or critical comments about one's appearance from parents, peers, or other significant others. For example, one study found that body-related comments received in childhood predicted body esteem in adulthood [49].

The fear of interpersonal rejection leads to a high amount of stress for women. People's sensitivity to rejection based on their appearance within interpersonal contexts are named appearance-rejection sensitivity (RS). Highly sensitive people are self-conscious about how they look. They anxiously anticipate that other people would reject them for their appearance. Moreover, when they are rejected, they attribute their appearance as the reason of rejection. Appearance-RS strongly predicted disruptive and excessive body image concerns [50]. The more sensitive participants were to being rejected based on their appearance, the more likely they were to report thoughts and behaviors characterizing body dysmorphic symptoms, to view cosmetic surgery as acceptable for both social and intrapersonal reasons, and to consider having cosmetic surgery in the future [50]. Appearance-RS also predicted social reasons for having cosmetic surgery [50]. This finding is consistent with research linking higher sensitivity with sociocultural influences, such as peer acceptance and feeling pressured to please others [37, 51, 52].

Women view their body image as an area that they can improve in order to gain social acceptance. In an experiment examining the implicit relations between rejection and appearance, female participants attempted to modify their body image in order to achieve self-enhancement [53]. Self-enhancement is one of two general sets of motives. When people are under threat (e.g., rejection, negative feedback, low self-esteem, depression, illness), their affective system and desires for self-enhancement are invoked [54]. Kunda proposed that the motivation to self-enhancement leads people to believe that they possess the desired traits, which in this case is smaller body size [55]. In another study, women who received self-esteem threats reported greater satisfaction with their appearance and less preoccupied with it than women who received positive feedback [56]. While the results indicate that participants held self-defensive perspectives immediately after receiving negative feedback, they might experience a paradoxical increase in investment in body image later on [57, 58]. As counterintuitive as this may seem, evidence shows that, after rejection and disapproval, people are motivated to protect their self-image by regulating their body image in order to maintain a balanced self-concept and diffuse the unbearable emotional distress. Compared to other aspects of the self, such as intelligence, one's body seems to be salient yet more malleable.

6. The crazy efforts in controlling body image

As a result of the interplay between sociocultural influences and the internalization process, women experience dissatisfaction towards their appearance. On the perceptual level, they are likely to overestimate their weight and size and see themselves as bigger than their actual size. Such irrational and inaccurate perception could lead to greater distress and a stronger motive to change. Consequently, individuals may develop distorted beliefs and automatic thoughts about their appearance and its significance [59].

As a result of the negative self-schema, individuals may resort to two common approaches to help regulate their body image. One is the avoidance approach, as some individuals are prone to avoid situations that might generate body image distress. For example, they might wear baggy clothes, avoid tight-fitting or revealing clothes, avoid mirror, and voluntarily isolate from social situations [60]. Other individuals may be prone to actively pursue the ideal body image and try to minimize the actual-ideal discrepancies [61]. They might monitor the condition of their body through repeated weighing and mirror checking. They are also preoccupied with their appearance and spend time-consuming efforts to groom and manage specific body areas. The more dangerous forms of effort include extreme restraints of eating behaviors and cosmetic surgeries. Regardless of the differences in approach, these actions are inherently self-reinforcing. They might relieve the individuals of immediate distress, but perpetuate the problems in the long term.

7. Conclusion

Struggles with body image have accompanied men and women throughout history and across cultures. With the development of research, we are fortunate to uncover the underlying mechanisms and pathways that link sociocultural influences and individual risk factors. However, many aspects of this problem still remain unresolved. Despite the active effort of studying and intervention, many individuals are still dissatisfied with their appearance and resort to alter their body in dysfunctional manners.

As we are embracing increasing diversity within our culture, it is imperative to reevaluate the dichotomous nature of our societal standards. When we put a category of qualities on the pedestal, we are essentially labeling people who do not possess such qualities as inadequate. Subsequently, individuals are subject to social scrutiny and risk disapproval from their significant and/or desirable others.

The stakes are raised obviously too high to the degree that it may erode individuals' self-esteem and overall well-being. In order for individuals to regain control over their body image, the real effort that we should spend is to challenge media messages and commonly held beliefs about beauty and help individuals develop a secure and stable self-concept that captures their true essence.

Author details

Wenting Mu[1*] and Fan Wu[2]

1 Center for the Treatment and Study of Anxiety, Philadelphia, US

2 Parent Planet Parenting Support Center, Nanjing, China

*Address all correspondence to: wenting.mu@pennmedicine.upenn.edu

References

[1] Cash, T.F. and Fleming, E.C. The impact of body image experiences: Development of the body image quality of life inventory. Int. J. Eat. Disorder. 2002;31:455-460. DOI:10.1002/eat.10033

[2] Cash TF. Body image: past, present, and future. Body Image. 2004;1(1):1-5. DOI:10.1016/S1740-1445(03)00011-1

[3] Cash TF, Deagle EA 3rd. The nature and extent of body-image disturbances in anorexia nervosa and bulimia nervosa: a meta-analysis. Int J Eat Disorder. 1997;22(2):107-125.

[4] Plato. Plato in Twelve Volumes, 250c., Vol. 9 translated by Harold N. Fowler. Cambridge, MA, Harvard University Press; London, William Heinemann Ltd. 1925.

[5] Mond, J. M., Mitchison, D., Latner, J. D., Hay, P. J., Owen, C., and Rodgers, B. Quality of life impairment associated with body dissatisfaction in a general population sample of women. BMC Public Health. 2013;13:920. DOI: 10.1186/1471-2458-13-920

[6] Quick, V., Nansel, T. R., Liu, D., Lipsky, L. M., Due, P., and Iannotti, R. J. Body size perception and weight control in youth: 9-year international trends from 24 countries. Int. J. Obes. 2014;38: 988-994. DOI: 10.1038/ijo.2014.62

[7] Mohr, H. M., Rickmeyer, C., Hummel, D., Ernst, M., and Grabhorn, R. Altered visual adaptation to body shape in eating disorders: implications for body image distortion. Perception. 2016;45: 725-738. DOI: 10.1177/0301006616633385

[8] Neagu, Alexandra and Francisc I. Rainer. Body image: a theoretical framework. The Publishing House Anthropology of the Romanian Academy. 2015;17(1):29-38.

[9] Madsen, Bohon, Feusner. Visual processing in anorexia nervosa and body dysmorphic disorder: similarities, differences, and future research directions. Journal of Psychiatric Research. 2013;47(10):1483-1491. DOI: 10.1016/j.jpsychires.2013.06.003.

[10] Frederick, David & Sandhu, Gaganjyot & Morse, Patrick & Swami, Viren. Correlates of appearance and weight satisfaction in a U.S. National Sample: personality, attachment style, television viewing, self-esteem, and life satisfaction. body image. 2016;17:191-203. DOI:10.1016/j.bodyim.2016.04.001.

[11] Liossi, C. Appearance related concerns across the general and clinical populations. [thesis] London: City University, 2003.

[12] Smolak L. Body image development in childhood. In: Cash TF, Smolak L, editors. Body image: A handbook of science, practice and prevention. New York: Guilford Press. 2011; 67-75.

[13] Wood KC, Becker JA, Thompson JK. Body image dissatisfaction in preadolescent children. J Appl Dev Psychol. 1996;17(1):85-100.

[14] Smolak L, Levine MP. Body image in children. In Thompson JK, Smolak L, editors. Body image, eating disorders, and obesity in youth: Assessment, prevention, and treatment. Washington: American Psychological Association. 2001; 41-66.

[15] Kearney-Cooke, A., & Tieger, D. Body image disturbance and the development of eating disorders. In: L. Smolak & M. D. Levine. The Wiley Handbook of Eating Disorders. West Sussex, UK: Wiley. 2015; pp. 283-296.

[16] Tiggemann M. Body image across the adult life span: Stability and change. Body Image. 2004;1(1):29-41.

[17] Paxton SJ, Neumark-Sztainer D, Hannan PJ, Eisenberg ME. Body dissatisfaction prospectively predicts depressive mood and low self-esteem in adolescent girls and boys. J Clin Child Adolesc Psychol. 2006;35(4):539-49.

[18] Cash TF, Fleming EC. Body image and social relations. In: Cash TF, Pruzinsky T, editors. Body image: A handbook of theory, research, and clinical practice. New York: Guilford Press. 2002; 277-86.

[19] Grogan S. Body image and health: Contemporary perspectives. J Health Psychol. 2006;11(4):523-30.

[20] Levine MP, Piran N. The role of body image in the prevention of eating disorders. Body Image. 2004;1(1):57-70.

[21] Littleton H, Breitkopf CR, Berenson A. Body image and risky sexual behaviors: An investigation in a tri-ethnic sample. Body Image. 2005;2(2):193-8.

[22] Wiseman CV, Turco RM, Sunday SR, Halmi KA. Smoking and body image concerns in adolescent girls. Int J Eat Disord. 1998;24(4):429-33.

[23] Cooley E, Toray T. Body image and personality predictors of eating disorder symptoms during the college years. Int J Eat Disord. 2001;30(1):28-36.

[24] Fairburn CG, Garner DM. The diagnosis of bulimia nervosa. Int J Eat Disord. 1986;5(3):403-19.

[25] The National Institute of Mental Health. Eating Disorders. 2016. Available from: https://www.nimh.nih.gov/health/topics/eating-disorders/index.shtml [Accessed: 2020-06-07]

[26] Bryn Austin. A call for the CDC to track eating disorders. 2017. Available from: https://www.hsph.harvard.edu/news/features/cdc-eating-disorders-tracking/ [Accessed: 2020-6-9]

[27] Bucchianeri, MM & Neumark-Sztainer, D. Body dissatisfaction: An overlooked public health concern. Journal of Public Mental Health. 2014;13(2),64-69. DOI:10.1108/JPMH-11-2013-0071

[28] "Beauty." Merriam-Webster.com Dictionary, Merriam-Webster. Available from: https://www.merriam-webster.com/dictionary/beauty.[Accessed: 2020-6-9]

[29] Cash, T.F. The psychology of physical appearance: Aesthetics, attributes, and images. In: Cash, T.F., & Pruzinsky, T.Body Images: Development, Deviance, and Change. The Guilford Press: New York; 1990. p. 51-7.

[30] Banaji, Mahzarin R., and Deborah A. Prentice. The Self in Social Contexts. Annual Review of Psychology. 1994;45(1): 297-332. DOI:10.1146/annurev.ps.45.020194.001501.

[31] Fallon, April. Culture in the mirror: Sociocultural determinants of body image. In: Cash, T.F. and Pruzinstky, T., Eds., Body Images: Development, Deviance, and Change. Guilford Press: New York; 1990. p.80-109. DOI:10.1002/erv.2400010112

[32] Czepczor-Bernat, Kamila & Kościcka, Katarzyna & Gebauer, Rafał & Brytek-Matera, Anna. Ideal body stereotype internalization and sociocultural attitudes towards appearance: a preliminary cross-national comparison between Czech, Polish and American women. Archives of Psychiatry and Psychotherapy. 2017;19: 57-65. DOI:10.12740/APP/78172.

[33] Van den Berg P, Thompson JK, Obremski-Brandon K, Coovert M. The Tripartite Influence model of body image and eating disturbance: a covariance structure modeling investigation testing the mediational

role of appearance comparison. J Psychosom Res. 2002;53(5):1007-1020. DOI:10.1016/s0022-3999(02)00499-3

[34] Thompson, J.K. and Heinberg, L.J. The media's influence on body image disturbance and eating disorders: we've reviled them, now can we rehabilitate them?. Journal of Social Issues. 1999;55:339-353. doi:10.1111/0022-4537.00119

[35] Nichter, Mark & Nichter, Mimi. Hype and weight. Medical Anthropology. 1991;13: 249-284. DOI: 10.1080/01459740.1991.9966051

[36] Kerry S. O'Brien et al. The relationship between weight stigma and eating behavior is explained by weight bias internalization and psychological distress. Appetite. 2015;102:70-76. DOI: 10.1016/j.appet.2016.02.032.

[37] Stice E. Sociocultural influences on body image and eating disturbance. In: Fairburn C.G and Bronwell K.D. Eating Disorders and Obesity: A Comprehensive Handbook. The Guildford Press: New York; 2002. p. 103-107.

[38] Lenny R. Vartanian, Shanta Dey. Self-concept clarity, thin-ideal internalization, and appearance-related social comparison as predictors of body dissatisfaction. Body Image. 2013;10(4):495-500. DOI:10.1016/j.bodyim.2013.05.004

[39] Kalodner C.R. Media influences on male and female non-eating-disordered college students: A significant issue, Eating Disorders. 1997;5(1): 47-57. DOI: 10.1080/10640269708249203

[40] Hamilton K, Waller *G. media* Influences on Body Size Estimation in Anorexia and Bulimia: An Experimental Study. British Journal of Psychiatry. 1993;162(6):837-840. doi:10.1192/bjp.162.6.837

[41] Irving, Lori M. Mirror images: effects of the standard of beauty on the self- and body-esteem of women exhibiting varying levels of bulimic symptoms. Journal of Social and Clinical Psychology. 1990;9(2):230-242.

[42] Pliner, P., Chaikin, S., & Flett, G. L. Gender differences in concern with body weight and physical appearance over the life span. Personality and Social Psychology Bulletin. 1990;16:263-273. DOI:10.1177/0146167290162007

[43] Strauman, Timothy J. et al. Self-discrepancies and vulnerability to body dissatisfaction and disordered eating. Journal of Personality and Social Psychology. 1991;61(6): 946-956. DOI: 10.1037/0022-3514.61.6.946

[44] Tiggemann, M., Martins, Y. and Churchett, L. Beyond Muscles: Unexplored Parts of Men's Body Image, Journal of Health Psychology. 2008;13(8): 1163-1172. DOI: 10.1177/1359105308095971.

[45] Swann WB, Jr, Stein-Seroussi A, Giesler RB. Why people self-verify. Journal of Personality and Social Psychology. 1992;62:392-401. DOI: 10.1037/0022-3514.62.3.392.

[46] Swann WB, Jr, Wenzlaff RM, Krull DS, Pelham BW. Allure of negative feedback: Self-verification strivings among depressed persons. Journal of Abnormal Psychology. 1992;101:293-306. DOI: 10.1037/0021-843X.101.2.293

[47] Levine MP, Piran N, Stoddard C. Mission more probable: Media literacy, activism, and advocacy as primary prevention. In: Piran N, Levine MP, Steiner-Adair C, editors. Preventing Eating Disorders: A Handbook of Inteventions and Special Challenges. Philadelphia, PA: Brunner/Mazel; 1999.

[48] de Vries DA, Vossen HGM, van der Kolk-van der Boom P. Social media and body dissatisfaction: investigating

the attenuating role of positive parent-adolescent relationships. J Youth Adolesc. 2019;48(3):527-536. DOI:10.1007/s10964-018-0956-9

[49] Lindsay Mclaren, Diana Kuh, Rebecca Hardy & Lise Gauvin. Positive and negative body-related comments and their relationship with body dissatisfaction in middle-aged women. Psychology & Health. 2004;19(2):261-272. DOI:10.1080/0887044031000148246

[50] Rachel M. Calogero, et al. Predicting excessive body image concerns among British university students: The unique role of Appearance-based Rejection Sensitivity. Body Image. 2010;7(1):78-81. DOI: 10.1016/j.bodyim.2009.09.005.

[51] Weisbuch M, Sinclair SA, Skorinko JL, Eccleston CP. Self-esteem depends on the beholder: effects of a subtle social value cue. J Exp Soc Psychol. 2009;45(1):143-148. DOI:10.1016/j.jesp.2008.07.021

[52] Park, L. E., DiRaddo, A. M. and Calogero, R. M. Sociocultural Influence and Appearance-Based Rejection Sensitivity Among College Students. Psychology of Women Quarterly. 2009;33(1):108-119. doi: 10.1111/j.1471-6402.2008.01478.x

[53] Mu, W, Schoenleber, M. Castro, A. and Berenbaum, H Body image distortion following social rejection or acceptance cues. Journal of Social and Clinical Psychology. 2019;38: 395-408. DOI: 10.1521/jscp.2019.38.5.395.

[54] Dufner, M. et al. Self-enhancement and psychological adjustment: a meta-analytic review. Personality and Social Psychology Review. 2019;23(1):48-72. DOI: 10.1177/1088868318756467

[55] Kunda Z. The case for motivated reasoning. Psychol Bull. 1990;108(3):480-498. doi:10.1037/0033-2909.108.3.480

[56] Josée L. Jarry, Amy L. Kossert. Self-esteem threat combined with exposure to thin media images leads to body image compensatory self-enhancement. Body Image. 2007;4(1):39-50. DOI: 10.1016/j.bodyim.2006.12.003

[57] Koff, E. and Sangani, P. Effects of coping style and negative body image on eating disturbance. Int. J. Eat. Disord. 1997;22:51-56.

[58] Robins, Richard & Beer, Jennifer. Positive illusions about the self: short-term benefits and long-term costs. Journal of Personality and Social Psychology. 2001;80:340-352. DOI:10.1037/0022-3514.80.2.340

[59] Jakatdar, Tejal A. et al. Assessment of body-image cognitive distortions. 2013. DOI:10.1037/t21197-000

[60] C. Alix Timko, et al. Body image avoidance: An under-explored yet important factor in the relationship between body image dissatisfaction and disordered eating. Journal of Contextual Behavioral Science. 2005;3(3):203-211. DOI:10.1016/j.jcbs.2014.01.002.

[61] Roz Shafran, Michelle Lee, Elizabeth Payne, Christopher G. Fairburn. An experimental analysis of body checking. Behaviour Research and Therapy. 2007;45(1):113-121. DOI:10.1016/j.brat.2006.01.015.

Chapter 12

Do Individuals with Eating Disorders See Their Own External and/or Internal Beauty?

Martha Peaslee Levine

Abstract

It has been well documented that individuals struggling with eating disorders don't have clear perceptions of their own bodies. Yet they overly rely on their body image as their sense of self. Even the criteria of certain eating disorders recognize that individuals are strongly affected by their body weight and shape, which is often seen through a distorted lens. Individuals with eating disorders, particularly anorexia nervosa, struggle not only with recognizing their external beauty but also their internal positive qualities. Their perfectionism and critical sense of self leads them to have negative views of their beauty and self-worth. This chapter will look at some of the reasons individuals with eating disorders struggle to appreciate their own beauty, internally as well as externally, and will offer some tools to help with these struggles. Many individuals, even those without disordered eating, struggle with critical self-perception. Perhaps this chapter can help us all become more compassionate to ourselves as we consider our external and internal aspects of beauty.

Keywords: Body image, eating disorders, perfectionism, beauty, self-identity, compassion

1. Introduction

Individuals with eating disorders often don't recognize their beauty. Those suffering from anorexia nervosa and bulimia nervosa struggle with accurate perceptions of their body weight and shape and unduly rely on body weight to define self-worth. [1] It is important to recognize that these individuals value themselves based on their weight and shape and yet often cannot accurately assess themselves in the physical dimension. I have worked with individuals in our eating disorders clinic who were emaciated but yet feared coming into a medical appointment because they would be weighed. The fear was not only about the number, which causes anxiety for many individuals with eating disorders, but a fear that they were so heavy that they would break the scale and/or exam table. One patient described that when he walked, he felt like he was breaking tiles because he was so heavy. He was actually severely underweight. Individuals with eating disorders struggle both in recognizing how they actually look and appreciating the beauty of their healthy bodies. They rely on their body weight to help define their self-worth and when they feel so negatively about how they look, they also struggle with seeing positives in themselves even their internal qualities.

When I have asked individuals in groups to share something that they like about themselves, they often describe that they like how they care about other people. They do not hold this same level of care or compassion for themselves. Individuals, particularly those with anorexia nervosa, can struggle with "repressed narcissism" such that they put other peoples' needs before their own. This can lead them to completely disregard their own needs, wants and desires. They see themselves as not worthy of being taken into consideration. These negative views can become intertwined. They see their body in a negative light and yet rely on their weight and body shape to define their self-worth. This can make them feel worthless and that negative feeling can contribute to them being even more self-critical about their worth and their body.

Additional challenging traits, such as perfectionism, are often associated with eating disorders and can further undermine a person's acceptance of his/her uniqueness and value. Perfectionistic tendencies makes it difficult for many individuals with eating disorders to complete certain tasks (such as sending emails) but also makes it hard for them to acknowledge their positive traits. They are often are so self-critical that they can never match their internal expectations. When asked in groups how they would define another person's success, the answer would often be, that the person would be successful if she tried her best. So if in school she got a "B" or a "C" but had tried her hardest, that person should still be seen as a success. However if they themselves didn't achieve a 100 percent on an assignment, then they saw themselves as worthless. Often there was a vast difference between their expectations of others and their expectations of themselves. Studies have linked disordered eating with body dissatisfaction, low self-esteem and perfectionism, especially in girls. [2] They struggle not only with dissatisfaction of their external beauty but struggle with their internal acceptance of self. The do not see their external or internal beauty.

Another chapter in this text, *Blossoming for Whom? Social Approval and Body Image*, looks at the role of social acceptance and body image disturbances in individuals struggling with eating disorders. [3] When individuals are faced with images and expectations that are not achievable, their own sense of beauty can be affected. Models are not only much thinner than the average woman but hours of styling and then subsequent photo-shopping make media images unattainable. If one bases her sense of value on external beauty standards, then it is understandable why she struggles with her sense of self when measured against these unrealistic expectations. If individuals struggling with eating disorders, do not see themselves in a realistic way, measure themselves against unreachable goals and link their self-value to these external perceptions, then it is understandable why they underestimate both their external and internal beauty. In addition, when individuals struggle with self-criticism and perfectionistic expectations, they often do not recognize their internal positives and by extension the beauty of their soul.

Are there ways to help individuals who struggle with eating disorders challenge these negative views? Can any of these tools help others who might not struggle with an eating disorder but who also struggle with low self-esteem and a negative view of self? My clinical roles include not only work within an eating disorders clinic but also providing therapy to students at Penn State College of Medicine. It is clear from my work with these individuals that although many do not struggle with eating disorders, they often struggle with a negative sense of self. Many describe elements of "the imposter syndrome", which affects many individuals even those functioning at a very successful level. They worry that they will be found out—that even though they have been successful and have received praise, it is only because they are fooling other people. When others discover who they really are—they will be seen as an imposter. How can any of us accept ourselves and discover our external and internal beauty? It seems like an impossible task considering all of the expectations that are placed on us. For individuals who struggle with eating disorders this journey can be even more difficult. Yet part of recovery will be working to discover and accept their own unique beauty.

While eating disorders and negative body image can affect all individuals, including men, my examples will be mostly from women that I work with in my clinic. They make up much of my treatment group. Some studies have suggested that body image disturbances and subsequent eating disorders occur at higher rates in groups of individuals who are trying to develop relationships with men because these women or gay men are often judged by their physical attractiveness. [4] In many societies, especially Westernized society, women are often evaluated and judged by their physical appearance. They are often exposed to comments or actions that focus only on their external appearance and objectify them as sexual possessions. Exposure to sexually objectifying comments can have a significant negative influence on women, not only when those comments are made directly but even when they are experienced secondhand. [5] It is for these reasons that this chapter will mainly focus on the experience of women, as they battle their illnesses, their negative self-perceptions and negative self-evaluations. They often don't see their external or internal beauty. Perhaps, there are ways to help them discover and accept their positive traits.

2. Body image

Body image is "a person's mental picture of how good or bad their physical appearance is, especially when compared with how they think they should look". [6] Within this simple definition, one can already see part of the difficulty because it focuses on how we think we should look. A comparison is introduced even in the definition. Comparisons can exacerbate negative perceptions, especially when comparing oneself to unobtainable images and struggling to accurately perceive oneself. "Your body image is what you think and how you feel when you look in the mirror or when you picture yourself in your mind. This includes how you feel about your appearance; what you think about your body itself, such as your height and weight; and how you feel within your own skin." [7] Body image encompasses many elements. Individuals who struggle with eating disorders, particularly anorexia nervosa, can have challenges with the perception of their bodies when they look in the mirror or see pictures of their own images. They not only feel negatively about themselves but can also struggle to perceive their bodies accurately.

When women struggling with an eating disorder were presented with an image of themselves and asked to photo-shop the image to what they believed they looked like, they routinely distorted portions of their body making themselves appear much heavier. This was opposed to a control group who made minimal adjustments to the images and accepted the photograph as an actual representation of how they looked. [8] Other studies indicate that individuals with anorexia nervosa might not overestimate their body size but that the disturbance comes because of their distorted view of a desired body. [9] This was further substantiated by [8]. When individuals with eating disorders photo-shopped their bodies to a desired state, these images were often extremely distorted and demonstrated unrealistic body goals. One study [10] suggests that individuals without an eating disorder often rate themselves as more attractive than others rate them. Whether this is true for the majority of individuals is not clear. It does seem, though, that individuals who do not struggle with an eating disorder often have a more compassionate relationship with their bodies. They recognize what their bodies do for them rather than being disconnected from their bodies. An individual with anorexia nervosa has disturbances that affect the person's functioning such that it is difficult to experience his/her body as an integrated aspect of self or develop a coherent narrative over time. [11] Clinically, individuals with anorexia nervosa often seem at war with their bodies, often seeing them as something that can be manipulated and brought under control even as these actions are taking them further and further away from being able to function in life. For women with bulimia, when they were asked to focus on their bodily sensations and estimate the width of their body, they overestimated their size by 13%, while the controls' estimates of how wide their bodies "felt" corresponded to actual sizes. [12] A positive body image is associated with psychological well-being. [13] It is clear that we need to consider how individuals can come to love their bodies and their selves.

3. Body dysmorphic disorder and obsessive compulsive disorder

Anorexia Nervosa and Body Dysmorphic Disorder (BDD) appear to share certain characteristics, such as a distorted and dissatisfied perception of one's body. The similarities and differences, i.e. how they are classified can be important to consider. As [14] notes, although they are classified differently—under feeding/eating disorders for anorexia nervosa and obsessive compulsive disorders for body dysmorphic disorder—their similarities may warrant classification as "body image disorders" and benefit from therapies that target body image disturbances. This overlap is important to consider as is the role that obsessive compulsive symptoms may play within eating disorders, specifically anorexia nervosa. Clinically, a number of patients whom I have worked with who struggle with anorexia nervosa often demonstrated obsessive compulsive tendencies. Typically their obsessive–compulsive behaviors revolved around food—not wanting to mix foods, needing to eat foods in a certain order, requiring a certain number of chews per bite, calculating and recalculating calories for the day to ensure that the total is correct. The intertwining of obsessive–compulsive behaviors and anorexia nervosa is interesting when considering a study that looked at emotional intuitiveness for individuals struggling with OCD (obsessive compulsive disorder). The authors [15] found that people with high obsessive–compulsive tendencies had more difficulty accessing internal states because of their continual doubting. Constant monitoring of thoughts can use up cognitive resources that can then make it difficult for individuals with OCD to have the resources available for actually experiencing the emotional states. [15] As individuals get more distant from their internal states, they start to doubt their own experiences or feelings and this can develop into a negative cycle. This is important to consider because individuals with anorexia nervosa often seem disconnected from their emotions. Often they use the illness as a mode of control and as a way to numb away feelings of discomfort. In [15] they suggest that many emotions do not stand up to constant monitoring but anxiety can "survive" this constant assessment and may even increase. These connections underscore some of the challenges for individuals with eating disorders as they try to recognize both their external and internal beauty. Individuals with anorexia nervosa share characteristics with BDD and as such cannot accurately assess their body and their external beauty. They often see it in a distorted perspective. The significant OCD tendencies that we see in individuals struggling with AN can affect their assessment of self. The disconnection with their emotions and the constant swirl of anxiety can make them doubt much of their inner life and challenge their recognition of their own emotional states and inner beauty.

4. Genetics

There are many ways that genetics can play a role in the perception of external and internal beauty in individuals with eating disorders, especially anorexia nervosa. One aspect relates to the genes that determine body shape and size. I have worked with young adolescent girls who were caught up in very negative dieting and other weight control behaviors as they struggled against the set-point of their bodies. Looking at their parents, it was clear that they had genes that kept them shorter and heavier than they desired. While they would acknowledge that they couldn't control or affect their height, they believed that their weight could be manipulated and controlled. This is a belief that most people seem to share. There is truth in this up to a point. If we eat more calories than we burn, we will gain weight. However, we each have certain physical characteristics that are determined by our genes. Some individuals will be taller or shorter, thinner or heavier, large bosomed or have a smaller chest size. One challenge is that much of Western society is weight-biased and so the messages that are given to individuals affects their ability to see their larger body as beautiful. When weight bias is internalized, it can lead to a lower quality of life, such as lower self-esteem. [16] This then affects individuals' "inner beauty" as they feel less positive about themselves.

Genetics also affects temperament and can influence how individuals feel about themselves. Individuals with restrictive eating patterns and a low body mass index demonstrated a repressed form of narcissism, such that they put others' needs before their own. [17] This is important to consider. How can one recognize her own value, self-worth and thereby inner beauty, if she is always sacrificing her needs to the needs of others? This brings us back to the earlier observation about the groups that I ran when I asked individuals what they liked about themselves. They often have a hard time identifying anything positive.

5. Influence of family

Family interactions can perpetuate eating disorder behaviors and negative attitudes towards oneself, especially related to negative comments offered by other family members. Family has the most powerful influence on values and norms related to appearance. [18] Clinically, I have seen many times how a mother's focus on her body dissatisfaction influences a daughter's body image dissatisfaction. One young woman was struggling to recover from severe anorexia nervosa and had to deal with her mother constantly asking if certain clothes made her (the mother) look fat or the mother would quiz the daughter about the number of calories in a certain food because she knew her daughter had all of that information present in her mind. It was hard for my patient to disengage from her eating disorder or not worry about her body image when she was constantly fielding these questions from her mother.

While modelling behavior can influence individuals' body image and body dissatisfaction, evidence suggests that parental comments and active encouragement to diet can have an even greater influence on their child's body concerns and eating behaviors. [19] Types of comments can vary with mothers focusing on health but dads and siblings making more negative comments and participating in teasing. [20] Positive feedback can improve self-esteem of individuals with eating disorders. [21] This is important for providers to remember because often our patients are attuned to negative feedback. We need to ensure that we approach them with an encouraging attitude. Too many times, clinicians do not consider the impact that their words will have on their patients. I have worked with clients who started down the road of an eating disorder because they were pushed to diet by a physician. We need to remember to focus on health and recognize that health can come at different sizes. Often physicians parrot the weight bias that is present in Westernized society. We need to work with families where individuals have been on the receiving end of negative comments and teasing and encourage the families to make a shift to neutral or positive comments. This can be hard but needs to be practiced. The family—especially parents—need to understand the negative effect that their comments can have on their child's self-esteem. [22]

6. Influence of peers

Many clients that I see in the eating disorders clinic describe weight-related teasing prior to the development of their eating disorder. The literature does support the fact that weight-related comments lead to body dissatisfaction and negative weight control behaviors. One study looked at the long-term effect of weight related bullying and examined whether there was a difference if the bullying occurred in the family or from peers. The authors followed a group of adolescents for 15 years and found that weight-based teasing in adolescence was associated with negative weight control actions even as adults. This included eating as a coping strategy, more body dissatisfaction, higher BMI and obesity risk. [23] Within this study, the authors found that the source of teasing (whether family or peers) had the same negative effect on girls but peer teasing had an increased effect on boys. [23] They were not certain whether girls were subject to more weight-related teasing than boys within their families or whether sensitivity to societal expectations related to weight and body image made girls more susceptible to the effects of teasing. Within their study, teasing also occurred irrespective of whether the individuals were overweight, obese or even underweight. In [24], the authors found that overweight or obese children who experienced teasing by their peers struggled with more depressive symptoms and also participated in more unhealthy weight control behaviors as a result of this teasing. It is clear that weight-related teasing by peers can have significant and long-lasting effects on individuals.

Another factor related to peers is that individuals with eating disorders often interpret ambiguous social interactions negatively. For example, when girls were asked to interpret social vignettes, those in the eating disorder group tended to attribute more hostile intent. For example, when given the situation that your best friend didn't invite you to their birthday party, the control group made the assumption that the friend, of course, expected her to come and advocated for checking out the issue by talking with the friend. Individuals with eating disorders, on the other hand, rarely picked the option of talking to the friend and made negative assumptions of the situation—that the friend had only been pretending to like her. [25] These social situations would often propel the patient group to consider using eating disorder behaviors rather than trying active ways to evaluate their negative assumptions. These concerns led to more anxiety in social situations and potentially more isolation from peers. To deal with these feelings of rejection, individuals often turn to eating disorder behaviors. This influences their sense of self and their self-esteem. Often they feel that the only thing that they have in their life that can consistently make them feel better is their eating disorder.

When considering self-esteem, we need to consider how it develops. Contingent self-esteem describes the relationship between receiving social approval and positive perception of oneself. Essentially one's self-esteem and positive feeling about oneself is dependent upon other's positive views of him/her and their approval.

This can lead to great concerns about one's weight, which can then influence body surveillance, body dissatisfaction, lower self-esteem and eating disorder behaviors. [26, 27] We need to consider how we help individuals develop their own positive self-assessment rather than always looking for social approval. That does not mean that we want people to not care at all or to try and alienate others. It does mean, though, that we have to help them recognize that we can't please everyone and that we need to feel okay about ourselves even if we meet with rejection in some social settings. In a recent session, a young woman, who suffers from an eating disorder, was struggling with how to handle her thoughts that others were always criticizing her or noticing if she gained a few pounds. She describes that she tried to handle her self-criticism by thinking why wouldn't everyone want to date her. This is a fallacy as well. We need to recognize black and white thinking and the potential sticky beliefs we can get trapped in. Just as not everyone is rejecting her because of her "perceived" flaws; she shouldn't expect everyone to overly accept her. The corollary from that would be that if someone rejects her then it is based on some aspect of her, rather than just the idea that we aren't going to mesh with everyone. I asked her if she wanted to date everyone. She doesn't. There are great people out there that for different reasons, we don't want to develop relationships with them. That does not mean that there is anything wrong with them or with us.

7. Other influences

7.1 Teachers, coaches and school policies

Many others can influence how an individual perceives him/herself. In schools, more extensive weight-related teasing leads to lower self-esteem and greater body dissatisfaction in girls. [28] We need to establish safety for individuals in families and schools so that they aren't continually subjected to negative comments. That means ensuring that teachers, nurses, administrators and coaches, for example, not only monitor their own comments and ensure that they are not making weight-based comments, but also ensure that weight-based bullying is squashed. Often weight-related teasing is not included in anti-bullying policy initiatives. [29] Yet it is an important challenge. There are ways that schools are possibly perpetuating weight focus and negative weight related behaviors. While one study suggested that weight-based teasing did not increase in schools even after introduction of obesity policies, such as measuring BMI and removing vending machines from schools, the teasing still stayed at 14% for overweight to obese children. [30] Since weight-based teasing can lead to negative health behaviors that can extend even into adulthood, it is clear that we need to continue to assess the effects of any interventions. Others studies have suggested that BMI measurements in schools can potentially lead to more eating disorder behaviors, especially if not handled correctly. [31] The authors point out that the CDC did not find enough evidence to recommend BMI report cards from schools. The challenge is that BMI is one number that if reported to parents can lead to a focus on weight and dieting. A higher BMI, though, could be related to a number of situations—an impending growth spurt or increased musculature from sporting activities. If families receive information about the BMI without guidance about what it means, more unhealthy weight control behaviors could occur, which can ultimately lead to more eating disorders or future obesity. [32] One thing to consider if we are working to try and improve overall health is whether eating disorder screening should occur in schools. In [33], we find that almost 15% of girls and 4% of boys score as having a possible eating disorder using the EAT-26. Since eating disorders have the highest mortality of any psychiatric disorders, strategies for early assessment and intervention would be useful. The SCOFF questionnaire has been suggested as a tool that could be used by school nurses to help identify children at risk for eating disorders. [34] As schools work to improve overall health and address concerns related to weight, this might be one assessment that should be implemented. With the current pandemic, the need for mental health assessment and interventions is being identified as being even more important. [35] It is incumbent on those in position of power/influence to consider the effect of their attitudes and words. Coaches can impact their athletes' view of their bodies and either inadvertently or on purpose push athletes towards negative weight control behaviors. The authors in [36] describe not only comments made to them that indicated the views and assessments of their bodies by coaches and those in power but also the objectification by others. I have worked with many young women who were encouraged to lose weight so they could be more competitive, could run faster, would look better, etc. Mary Cain in a NYTimes editorial described how she was influenced by her coaches to get thinner and thinner until her body broke down. [37] One study found that while coaches did have an understanding of eating disorders, they, at times, didn't understand the significance of some symptoms, such as amenorrhea. [38] The authors also found that coaches preferred to talk with teammates if they had a concern about an athlete rather than turning to specialists in the field of eating disorders. [38] It will be important to ensure that coaches are familiar with aspects of eating disorders and recognize the need to get athletes the care that they need.

7.2 Media

The effect of television and Westernization has been established through many studies. When television was introduced to Fiji, body image dissatisfaction and disordered eating behaviors increased. [39] This was replicated in Bhutan, which was the last country exposed to television. When this occurred, it led to a subsequent pressure to get thinner, look thinner and regulate one's appearance based on media images rather than compared to one's peer group. [40] Media has a strong influence, not only on distorted perceptions and expectations, but on how role models are perceived. Intense celebrity worship by young teens can have a negative influence on their body image. [18, 41] It isn't clear which comes first—does obsession with a celebrity lead to poor body image or do individuals with poor body image gravitate towards celebrity fixation. Interestingly one study looked at body dissatisfaction, restrained eating and compared congenitally blind women, acquired blind women and sighted women. They found that these issues increased related to the extent of visual media exposure so that congenitally blind women had overall better body image because they had not been faced with the media's distorted expectations. [42] In the tools, we will discuss the importance of media literacy which can help with body dissatisfaction. One program, ARMED, which offered a two session media literacy course demonstrated that 8 weeks after participating in the course, women who were at high risk for developing an eating disorder had less body dissatisfaction. A change was not noted for women at low risk for developing an eating disorder but they may be less influenced in their sense of self and their body image as compared to those at higher risk. [43]

7.3 Objectification

"Objectification theory posits that girls and women are typically acculturated to internalize an observer's perspective as a primary view of their physical selves." ([44], p.173) This can lead to habitual body monitoring, which increases shame and anxiety for many women. Seeing oneself as an external viewer of one's body can also decrease awareness of internal body states. [44] We have seen that the connection with one's body and reliable interpretation of internal body states can be affected for individuals who struggle with eating disorders. They often monitor their bodies and perform numerous body checking activities to evaluate how they should be feeling about their body. I worked with one young woman who agreed to not weigh herself but turned instead to measuring and recording the size of many of her body parts—circumference of thighs, etc. As with the objectification theory—she was seeing herself as just the sum of her parts, which is how the authors discuss their theory. "The common thread running through all forms of sexual objectification is the experience of being treated as a body (or collection of body parts) valued predominantly for its use to (or consumption by) others." ([44], p. 174).

This objectification not only affects how they see and monitor their physical bodies but also how they function in their overall lives. "Again, though, the habitual self-conscious body monitoring that results from self-objectification might best be viewed as a strategy many women develop to help determine how other people will treat them, which has clear implications for their quality of life." ([44], p. 180) Because they recognize that they are being objectified by others, they constantly need to scan themselves to assess how they will be perceived. Rather than being in the moment and living their lives, they are viewing and measuring themselves. It has also been demonstrated that women do not have to be personally targeted to experience the negative effects of objectification. [5] One mediating factor appeared to be that when individuals feel part of a group, they can experience less negative self-esteem stemming from sexism. [45] Feeling part of a group can be a larger challenge for individuals struggling with eating disorders. As we discussed earlier, they may have more negative evaluations of interpersonal interactions, which can leaving them feeling not a part of a group and turning more to their eating disorder. [25] This can affect this described mediating factor.

This objectification can be even more difficult for young girls as they progress through puberty. "Far beyond the idea that adolescent girls simply do not like the size and shape of their maturing body, girls learn that this new body belongs less to them and more to others." ([44], p. 193) For many of my patients, going through puberty led to increased negative feelings about themselves. If they went through puberty earlier than others, they were often targeted for comments and felt self-conscious because their bodies looked different than others who were the same age. One patient described that when she went through puberty, her family made fun of her changing body—often it was her brothers but her parents did not intervene and on occasion, her father made comments. Now it is hard for her to gain weight; it is hard for her to develop curves or breasts. Others have commented about similar situations. After hearing her brother and father comment consistently on women's breasts, a young woman feels very ambivalent and self-conscious about gaining weight and developing the same physical attributes that garnered so much attention from the men in her life.

7.4 Trauma

Trauma can affect an individual and her sense of beauty and herself. It can be associated with disordered eating; this is often because of negative emotions and thoughts. [46] I have certainly worked with individuals who blamed themselves for the attacks. This led to not only negative feelings but also the use of disordered eating to try and gain back a sense of control or to numb the negative feelings related to the trauma. The authors [46] also see these effects of trauma. Disordered eating can be a way to avoid unwanted attention and the individual may feel this is a way to keep them safe. [46] One woman whom I worked with had recovered from her eating disorder but was then a victim of a sexual assault. She began her pattern of restricting again. She described to me that if she was assaulted when she looked "healthy" and had curves, she needed to make sure she didn't have that happen again. Others have described working to disappear. There is often an ambivalence—should I be thinner/smaller so that I am not noticed or since thinness is often associated with sexual objectification, will that draw more attention? It can put women in conflicted and negative associations with their bodies.

Emotional abuse contributes to negative self-perception, which then can contribute to disordered eating, including night eating syndrome and binge eating disorder. [47] One study found that adolescent girls who were exposed to traumatic life events or who had to navigate difficult family situations were at an elevated risk of becoming obese and in engaging in unhealthy weight control behaviors. [48] One study found that patients with bulimia were more likely to have a history of trauma as compared to other eating disorder diagnosis [49] What is even more significant from this study is that almost 35% of the adolescents being treated for eating disorders had a history of trauma. [49] A systematic review demonstrated an association between adverse life experiences and obesity and binge eating disorder. [50] Adults who have experienced one adverse childhood experience (ACE) were likely to have suffered multiple other adverse experiences during childhood. [51] This is important because ACES appear to have a cumulative effect such that as the number of events increases so does the effect on mental health. [52] It seems fairly obvious how abuse can undermine an individual's perception of herself. This can lead to the use of food or disordered eating to try and control negative emotions. It can also influence acceptance of self and recognition of both external and internal positive qualities. I have worked with fabulous young women who do not seem to recognize any of their positive attributes because of being criticized and belittled by other's in their lives—often family members. When working with clients with disordered eating, it is helpful to assess their perception of themselves and any history of emotional or other types of abuse.

8. Tools to help

A. In our eating disorder program, one therapist who runs a body image group presented at a conference with me [53] and outlined some exercises that she uses in groups to help individuals consider misconceptions or beliefs that they may harbor about their bodies. These tools also allow them to start to consider their body from the perspective of function—rather than objectification. As we have seen in this chapter, that can be a very important strategy. Some of the group ideas include:

- Consider your body's needs and rhythms. If someone was going to move into your body for the day, what important information would they need to know about it? What guidelines would you need to provide related to rest, feeding, watering, light/outdoors, physical activity, touching, soothing, healing--what keeps it functioning well?

- A letter From Your Body to Yourself. Our bodies hear everything we think/say about ourselves...if it could write a letter to you, what would it say?

- List three women/men you would like to be for a day and explain why it would be a great experience to be them. She finds that usually patients include individuals based on positives in their stories or their accomplishments and not often focused on appearance.

- Letter to younger child/self. What would you tell a younger child or your younger self to avoid an eating disorder?

- Body Checking reflection. What are you trying to find out and how does it help?

- My Body Experience from then to now. A time-line for influential events and experiences that influence your body image. This can help individuals understand defining moments and beliefs that they developed about themselves.

B. We have discussed trauma as it relates to eating disorders. It is very important when working with individuals to have a compassionate strategy when eliciting a trauma history. It can often help when assessing trauma to start with opened ended questions. One that I like to use: "Has anything happened in your life that you are still struggling with? That has been hard for you to get past?" This will often lead to stories of losses in their lives, or previous bullying or heart breaking stories of violent death. A few things to keep in mind when asking these questions is that you will need to provide the space and support for the patient to share his/her story. When getting a trauma history, I don't think that anything can be worse than being dismissive or uninterested. This will reinforce the sense that victims sometimes have that they are to blame for what happened. The other important thing is to be able to hold the difficult emotions that might come up in the session—both your reaction and that of your client. If someone shares a story of past trauma, they are demonstrating a belief that they can trust you. It is vital as a practitioner that you honor that trust. If the question that I shared above doesn't elicit a story or depending on what they say, I do go through each type of potential past trauma. "Has anyone ever made you feel bad about yourself?" This can elicit emotional abuse, but if you don't get an answer, you can ask about past bullying. As we discussed in this article, weight-related teasing/bullying can affect self-perception and lead to disordered eating. "Has anyone ever been physically harmful to you?" This is to obtain physical abuse. If in your history there has been any mention of substance abuse, you might want to add another question to ask that when so and so was drinking, did they become harmful to your or anyone else in your family? This will get their experience but also can elicit secondary trauma by witnessing abuse of someone else. "Are there any sexual situations that you have been put into that made you uncomfortable?" When getting this history, it is important to maintain eye contact so that the individual knows that you are interested. It is important to give them the space to answer. Do not group all types of abuse together and ask in such a way that it seems like it is just a box that needs to be checked off. I have seen individuals do that in evaluations—it doesn't open up the necessary space. It doesn't make them look interested. Also make sure that others, such as family members, aren't in the room when you are taking this history. How can you ask about potential abuse and expect to get the truth if someone else is in the room, listening to the story.

C. When treating individuals with eating disorders, especially anorexia nervosa, there is a need for increasing the patient's self-esteem and self-worth. [54] That does certainly seem to be a very important tool. It may be less clear how to accomplish this. One way would be to help patients recognize when they are having negative or critical thoughts about themselves. Sometimes this can occur so routinely that they don't even stop to examine their thoughts or they believe that everyone harbors this internal self-criticism. It is important to clarify the patients' thoughts and work to help them recognize them and challenge them. Is there another way to reframe the thought? For example, if someone is critical of themselves because of past abuse, can they instead view their resilience as a strength? Are there better ways to frame things—if they describe themselves as "stupid" because they didn't do well on an exam, can you help them put this in a reasonable context? We all make mistakes or can fail at something but that does not make us total failures. Are they over-generalizing? Are they turning one small setback into a catastrophe? One tool that I find helpful when faced with patients who harbor negative thoughts or self-criticism is to ask them if they would tell those same comments to a friend or family member. Typically they would not. The question then is why they feel that they should tell it to themselves. This can help them start to recognize when an interior monologue is helpful or when it is perpetuating a negative cycle.

D. Self-blame and self-distraction perpetuate an individual's negative assessment. [55] Coping strategies can help with body comparisons and body dissatisfaction such that positive reframing can help individuals challenge negative views. What is positive reframing? Again, it is shifting from the negative point of view of not being good enough to highlighting positives. "I might not be as thin as her but I'm healthy." "My body is strong and fit." "I am more than just my body. I am a unique and talented individual." Focusing on criticisms of ourselves can lead to more body dissatisfaction and overall self-despair. Studies have demonstrated that focusing more on body functionality can lead to more body positivity and self-acceptance. [56]

E. We can push our media to offer more realistic sized models. When individuals viewed images of full-figured women, they had improvement in their body appreciation. [57] Related to celebrities, especially those who offer thinspiration or appearance focused accounts, viewing these images can affect one's body-image and body-acceptance. However, it has been found that exposure to parodies of these accounts can lead to improved body satisfaction and mood. [58] So go, have a laugh, feel better about yourself. One of my favorites is Celeste Barber's recreation of other videos. [59]

F. Consider the impact of children's toys as well. Girls exposed to Barbie dolls demonstrated lower body esteem and wanted a thinner body shape as compared with controls. While older girls didn't demonstrate this same difference, there is a suggestion that exposure to dolls that foster an unrealistically thin body ideal may damage girls' body image. [60] If you look at how many toys, especially super heroes, have changed over the years, one can understand that these subliminal expectations are not just being presented to young girls but also to young boys who are being pushed to desire increased muscularity. We need to push manufacturers to create more appropriate styled toys or consider the toys that we provide the children in our lives.

G. An impressive video, which was designed to address the unrealistic expectations that the media has been foisting on women, is the Dove Evolution video. [61] You can find it on YouTube. It shows a beautiful young woman being transformed into a model on a billboard—including make-up, hair and photo-shopping. It reminds us that the expectations that we are comparing ourselves to are not realistic. Can this have an impact? Yes, it can. When presented to a group of young women who watched music videos, it was demonstrated that this simulated commercial break had the potential to disrupt the detrimental effects of social comparisons with the idealized models. [62] All young women should understand that when they encounter the media, they are comparing themselves to manipulated and often unobtainable images.

H. Consider the impact of social network use. This can be a challenge, especially now in light of the pandemic, because so much of our social connections are occurring on-line. What can be helpful to realize is that not all social network use is equal in the potential negative impact on body image. It appears to be more appearance focused activities that can have a negative impact. Focusing on photographs on Facebook rather than amount of Facebook activity appears to influence thin-idealism. On Instagram, again spending time on appearance-focused accounts rather than appearance neutral had more of an impact on negative body-image and self-worth. [63] Body positive imagery, focusing on diverse beauty and functionality, can change the focus and improve body acceptance. [64] In addition, positive parental support and a strong school environment that teaches media literacy has helped mitigate the potential negative influence of social media. [65] This is important to consider. Research has demonstrated the negative and or positive effects that school environments can have on young adults. If schools don't discourage weight-bullying, it has a negative impact. If, though, school officials recognize their role as positive influencers of our young adults and model positivity, media literacy, and appreciation of differences then they can have a constructive impact.

I. As per [18] parental spoken observations, both positive and negative, can have a significant impact on body-perception. Often observations and discussions of food and body can slip into conversations unnoticed—they have become so much a norm in many societies. Yet we all need to be aware of the potential our words can have on those around us, especially our children who are in the process of developing their sense of identity and self-worth. I wrote a previous chapter about communication challenges with suggestions of how to improve communication especially related to individuals struggling with eating disorders and low self-worth. [66] As discussed in family interactions, it is clear that different family members may interact differently—some in the guise of trying to help, some with critical comments or teasing. Different interventions may be needed depending on the ages or relationship of the individuals or the types of comments that are being made. [20]

J. Self-compassion can influence body-positivity, improve appearance self-esteem and buffer against risk factors. [67, 68] One must ask if there is any good argument against self-compassion. Many of our clients who struggle with eating disorders are very critical of themselves. If we can help them improve self-compassion, it may not only improve their positive feelings about themselves but could influence their willingness to care for their bodies.

K. There are programs that have been designed to help reduce drive for thinness and linking one's self-worth to the approval of others. *In Favor of Myself* is one such program that helps adolescents cope with the challenges of their life stages. [69] This is important to consider. Different life stages lead to different challenges and potential coping strategies. I often work with adolescents who are struggling with trying to navigate middle school and high school. The desire for social approval in light of shifting alliances and bullying make it challenging for everyone but in particular those individuals who link their sense of self-worth to the approval of others. If we can provide tools to help our adolescents navigate these unsettling years, it can help improve self-confidence and positive coping strategies. *Expand Your Horizon* is a program than can help focus on body functionality, which we have discussed can improve body appreciation and decrease disordered eating. [70]

L. One area that needs more study is body image directed techniques, such as mirror exposure, video confrontation, and virtual reality body exposure. These techniques may be helpful in improving body image outcomes but evidence is still unclear. [71]

M. One study [54] suggested that patients with anorexia nervosa may have a left-hemispheric bias related to their body distortions. They suggest working to activate the patient's right hemisphere by left-hand contractions or engaging the patient in EMDR (Eye Movement Desensitization and Reprocessing). This could be an interesting tool to include in future research.

N. Authors of [72] looked at movement from a negative body image to a positive sense of self during development from adolescence to young adulthood. Turning points included, finding a new and supportive social context, such as new friends or a romantic partner. This allowed individuals to have a feeling of belonging and acceptance. Another aspect was increased empowerment—excelling in certain sports or developing their unique style. Finally, they actively used strategies to practice body acceptance and valued themselves and their lives in larger contexts than just their body. From this study, we return to self-compassion again. It is important to help adolescents begin to consider how their body can help them lead a valued life rather than just focusing on its weight and shape.

9. Conclusion

It can be challenging for anyone to appreciate his/her external and internal beauty, but this is even more difficult for individuals who struggle with eating disorders. Photo-shopped media images bombard us and set up unobtainable expectations. Negative comments from others can affect body-image and self-esteem. Likewise attention that pushes individuals, especially women, to be seen as objects rather than whole vulnerable humans can affect sense of self and also perpetuate trauma. Healthcare practitioners and all humans need to consider how to encourage self-compassion and improve self-esteem. We need to help others see their own beauty and we need to recognize it in ourselves.

Conflict of interest

The author has no conflicts of interest.

Author details

Martha Peaslee Levine
Penn State College of Medicine, Hershey, Pennsylvania, United States of America

*Address all correspondence to: mpl12@psu.edu

References

[1] American Psychiatric Association. *Diagnostic and statistical manual of mental disorders, Fifth Edition.* (5th ed.). Washington, DC: APA Press. 2013. https://doi.org/10.1176/appi.books.9780890425596

[2] Teixeira MD, Pereira AT, Marques MV, Saraiva JM, de Macedo AF. Eating behaviors, body image, perfectionism, and self-esteem in a sample of Portuguese girls. Revista Brasileira de Psiquiatria. 2016; 38:135-140 doi:10.1590/1516-4446-2015-1723 00000000-0002-731

[3] Wenting M, Fan W. Blossoming for Whom? Social Approval and Body Image. In: Levine MP, Santos JS, editors. Beauty: Cosmetic Science, Cultural Issues, and Creative Developments. IntechOpen, Croatia: IntechOpen. DOI: 10.5772/intechopen.94503. Available from: https://www.intechopen.com/online-first/blossoming-for-whom-social-approval-and-body-image [Accessed 2020-07-12]

[4] Seiver MD. Sexual orientation and gender as factors in socioculturally acquired vulnerability to body dissatisfaction and eating disorders. Journal of Consulting and Clinical Psychology. 1994; 62 (2): 252 – 260.

[5] Koval P, Holland E, Zyphur MJ, Stratemeyer M, Knight JM, Bailen NH, Thompson R J, Roberts T-A, Haslam N. How does it feel to be treated like an object? Direct and indirect effects of exposure to sexual objectification on women's emotions in daily life. J Pers Soc Psychol. 2019 Jun; 116(6):885 – 898. Doi 10.1037/pspa0000161. Epub 2019 Apr 29.

[6] Oxford Learner's Dictionaries [Internet]. 2020. Available from: https://www.oxfordlearnersdictionaries.com/us/definition/english/body-image [Accessed 2020-10-31]

[7] Office on Women's Health [Internet]. 2019. Available from: https://www.womenshealth.gov/mental-health/body-image-and-mental-health/body-image [Accessed 2020-10-31]

[8] Nakhla M, Levine MP. "What I Think I Look Like": A photographic study of perceived body image in patient with eating disorders. Poster Presentation, ICED. 06-2017, Prague, Czech Republic.

[9] Mölbert SC, Thaler A, Mohler BJ, Streuber S, Romero J, Black MJ, Zipfel S, Karnath H-O, Giel KE. Assessing body image in anorexia nervosa using biometric self-avatars in virtual reality: Attitudinal components rather than visual body size estimation are distorted. Psychological Medicine. 2018; 48: 642 – 653. Doi: 10.1017/S0033291717002008

[10] Jansen A, Smeets T, Martijn C, Nederkoorn C. I see what you see: The lack of a self-serving body-image bias in eating disorders. British Journal of Clinical Psychology. 2006; 45: 123 – 135.

[11] Amianto F, Northoff G, Abbate Daga G, Fassino S and Tasca GA. Is anorexia nervosa a disorder of the self? A psychological approach. Front. Psychol. 2016; 7:849. doi: 10.3389/fpsyg.2016.00849

[12] Franzen U, Florin I, Schneider S, Meier M. Distorted body image in bulimic women. Journal of Psychocomatic Research. 1988; 32 (425): 445 – 450.

[13] Swamia V, Weisb L, Barronc D, Furnhamd A. Positive body image is positively associated with hedonic (emotional) and eudaimonic (psychological and social) well-being in British adults. The Journal of Social Psychology. 2018; 158 (5): 541 – 552. doi.org/10.1080/00224545.2017.1392278

[14] Phillipou A, Castle DJ, Rossell SL. Direct comparisons of anorexia nervosa

and body dysmorphic disorder: A systematic review. Psychiatry Research. 2019; 274: 129 – 137.

[15] Dar R, Lazarov A, Liberman N. How can I know what I'm feeling? Obsessive-compulsive tendencies and induced doubt are related to reduced access to emotional states. J Behav Ther & Exp Psychiat. 2016; 52: 128 – 137.

[16] Walsh OA, Wadden TA, Tronieri JS, Chao AM and Pearl RL. Weight bias internalization is negatively associated with weight-related quality of life in persons seeking weight loss. Front Psychol. 2018; 9:2576. doi: 10.3389/fpsyg.2018.02576

[17] Brunton JN, Lacey JH, Waller G. Narcissism and eating characteristics in young non-clinical women. Journal of Nervous and Mental Disease. 2005; 193(2): 140 - 143. ISSN 0022-3018 https://doi.org/10.1097/01.nmd.0000152784.01448.fe

[18] Heiman T, Olenik-Shemesh D. Perceived body appearance and eating habits: The voice of young and adult students attending higher education. International Journal of Environmental Research and Public Health. 2019; 16: 451. doi:10.3390/ijerph16030451

[19] Rodgers R, Chabrol H. Parental attitudes, body image disturbance and disordered eating amongst adolescents and young adults: A review. Eur Eat Disorders Rev. 2009; 17: 137 - 151.

[20] Berge JM, Hanson-Bradley C, Tate A, Neumark-Sztainer D. Do parents or siblings engage in more negative weight-based talk with children and what does it sound like? A mixed-methods study. Body Image. 2016; 18: 27-33

[21] Vanderlinden J, Kamphuis JH, Slagmolen C, Wigboldus D, Pieters G, Probst M. Be kind to your eating disorder patients: The impact of positive and negative feedback on explicit and implicit self-esteem of female patients with eating disorders. Eating Weight Disord. 2009; 14 (4): e237-e242.

[22] Rodgers RF, Paxton SJ, Chabro H. Effects of parental comments on body dissatisfaction and eating disturbance in young adults: A sociocultural model. Body Image. 2009; 6: 171 – 177.

[23] Puhl RM, Wall MM, Chen C, Austin SB, Eisenberg BE, Neumark-Sztainer D. Experiences of weight teasing in adolescence and weight-related outcomis in adulthood: A 15-year longitudinal study. Prev Med. 2017 July ; 100: 173-179. doi:10.1016/j.ypmed.2017.04.023.

[24] Madowitz J, Knatz S, Maginot t, Crow SJ, Boutelle KN. Teasing, depression and unhealthy weight control behaviour in obese children. Pediatric Obesity. 2012; 7: 446-452.

[25] McFillin RK, Cahn SC, Burks VS, Levine MP, Lane-Loney S, & Levine RL. Social information-processing and coping in adolescent females diagnosed with an eating disorder: toward a greater understanding of control. Eating Disorders. 2012; 20: 42-59. doi:10.1080/10640266.2012.635565

[26] Grossbard JR, Lee CM, Neighbors C, Larimer ME. Body image concerns and contingent self-esteem in male and female college students. Sex Roles. 2009; 60 (3-4): 198 – 207. Doi: 10.1007/s11199-008-9535-y

[27] Moya-Garófano A, Moya M. Focusing on one's own appearance leads to body shame in women but not men: The mediating role of body surveillance and appearance-contingent self-worth. Body Image. 2019; 29: 58 – 64.

[28] Lampard AM, MacLehose RF, Eisenberg ME, Neumark-Sztainer D, Davison KK. Weight-related teasing in the school environment: Associations with psychosocial health and weight

control practices among adolescent boys and girls. J Youth Adolescence. 2014; 43: 1770 – 1780. doi: 10.007/s10964-013-0086-3

[29] Puhl RM, Luedicke J, King KM. Public attitudes about different types of antibullying laws: Results from a national survey. Journal of Public Health Policy. 2015; 36: 95-109. doi:10.1057/jphp.2014.46;

[30] Krukowski RA, West DS, Siddiqui NJ, Bursac Z, Phillips MM, Raczynski JM. No change in weight-based teasing when school-based obesity policies are implemented. Arch Pediatr Adolesc Med. 2008; 162(10): 936-942. doi: 10.1001/archpedi.162.10.936.

[31] Ikeda JP, Crawford PB, Woodward-Lopez G. BMI screening in schools: helpful or harmful. Health Education Research. 2006; 21(6): 761-769.

[32] Siahpush M, Tibbits M, Shaikh RA, Singh GK, Kessler AS, Huang TTK. Dieting increases the likelikhood of subsequent obesity and BMI gain: Results from a prospective study of an Australian national sample. Int.J. Behav. Med. 2015; 22:662-671.DOI 10.1007/s12529-015-9463-5.

[33] Austin SB, Ziyadeh NJ, Forman S, Prokop LA, Keliher A, Jacobs D. Screening high school students for eating disorders: results of a national initiative. Prev Chronic Dis 2008;5(4). http://www.cdc.gov/pcd/issues/2008/oct/07_0164.htm. Accessed [*03/28/2021*].

[34] Rindahl K. A systematic review of literature on school screening for eating disorders. International Journal of Health Sciences.2017; 5(3):1-9. ISSN: 2372-5060 (Print), 2372-5079 (Online). DOI: 10.15640/ijhs.v5n3a1

[35] Marques de Miranda D, Athanasio BdaS, Oliveira ACS. How is COVID-19 pandemic impacting mental health of children and adolescents?International Journal of Disaster Risk Reduction *51.2020: 101845*.https://doi.org/10.1016/j.ijdrr.2020.101845. Accessed [03/28/2021]

[36] Stoll SK, VanMullem H, Ballestero N, Brown L. Beauty and the beast: Perception of beauty for the female athlete. Perception of Beauty. Intechopen. 2017. http://dx.doi.org/10.5772/intechopen.68946

[37] Cain M. [Internet] Available from: https://www.nytimes.com/2019/11/07/opinion/nike-running-mary-cain.html?action=click&module=RelatedLinks&pgtype=Article [Accessed on: 2020-11-1]

[38] Sherman RT, Thompson RA, Dehass D, Wilfert M. NCAA coaches survey: The role of the coach in the identifying and managing athletes with disordered eating. Eating Disorders. 2005; 13:447-466. DOI: 10.1080/10640260500296707

[39] Becker AE, Burwell RA, Gilman SE, Herzog DB, Hamburg P. Eating behaviours and attitudes following prolonged exposure to television among ethnic Fijian adolescent girls. British Journal of Psychiatry. 2002; 180: 509 – 514.

[40] Stern JM. Transcultural aspects of eating disorders and body image disturbance. Nordic Journal of Psychiatry. 2018; 72 (S1): S23–S26. doi.org/10.1080/08039488.2018.1525642

[41] Maltby J, Giles CD, Barber L, McCutcheon LE. Intense-personal celebrity worship and body image: Evidence of a link among female adolescents. British Journal of Health Psychology. 2005; 10: 17 – 32. doi: 10.1348/135910704X15257

[42] Ashikali E-M, Dittmar H. Body image and restrained eating in blind and

sighted women: A preliminary study. Body Image. 2010; 7: 172 – 175.

[43] Coughlin JW, Kalodner C. *Media* literacy as a prevention intervention for college women at low- or high-risk for eating disorders. Body Image. 2006; 3: 35-43

[44] Fredrickson BL, Roberts T-A. Objectification Theory: Toward understanding women's lived experiences and mental health risks. Psychology of Women Quarterly. 1997; 21: 173- 206.

[45] Spencer-Rodgers J, Major B, Forster D, Peng K. The power of affirming group values: Group affirmation buffers the self-esteem of women exposed to blatant sexism. Self Identity. 2016; 15(4): 413 – 431. Doi: 10.1080/15298868.2016.1145593

[46] Breland JY, Donalson R, Dinh JV, Maguen S. Trauma exposure and disordered eating: A qualitative study. Women Health. 2018; 58(2): 160 – 174. doi: 10.1080/03630242.2017.1282398

[47] Hymowitz G, Salwen J, Salis KL. A mediational model of obesity related disordered eating: The roles of childhood emotional abuse and self-perception. Eat Behav. 2017; 26: 27 – 32. doi: 10.1016/j.eatbeh.2016.12.010

[48] Isohookana R, Marttunen M, Hakko H, Riipinen P, Riala K. The impact of adverse childhood experiences on obesity and unhealthy weight control behaviors among adolescents. Comprehensive Psychiatry. 2016; 71: 17-24

[49] White AAH, Pratt KJ, Cottrill C. The relationship between trauma and weight status among adolescents in eating disorder treatment. Appetite. 2018; 129: 62-69

[50] Palmisano GL, Innamorati M, Vanderlinden J. Life adverse experiences in relation with obesity and binge eating disorder: A systematic review. Journal of Behavioral Addictions. 2016; 5(1): 11-31 DOI: 10.1556/2006.5.2016.018

[51] Dong M, Anda RF, Felitti VJ, Dube SR, Williamson DF, Thompson TJ, Loo CM, Giles WH. The interrelatedness of multiple forms of childhood abuse, neglect, and household dysfunction. Child Abuse & Neglect. 2004; 28: 771-784

[52] Cohen RA, Hitsman BL, Paul RH, Mccaffery J, Stroud L, Sweet L, Gunstad J, Niaura R. Macfarlane A, Bryant RB, Gordon E. Early life stress and adult emotional experience: An international perspective. Int'l J. Psychiatry in Medicine. 2006; 369(1): 35-52.

[53] Levine MP, Clark-Byers, B. Body image: What research and clients say and how to respond. Renfrew Conference Workshop, 2018-11-19, Philadelphia, PA, USA.

[54] Kazén M, Baumann N, Twenhöfel JF, Kuhl J. When do anorexic patients perceive their body as too fat? Aggravating and ameliorating factors. PLoS ONE 14(2): e0212612. doi. org/10.1371/journal.pone.0212612

[55] Pinkasavage E, Arigo D, Schumacher LM. Social comparison, negative body image, and disordered eating behavior: The moderating role of coping style. Eating Behaviors. 2015; 16: 72 – 77.

[56] Alleva JM, Gattario KH, Martijn C, Lunde C. What can my cody do vs. how does it look? : A qualitative analysis of young women and men's descriptions of their body functionality or physical appearance. Body Image. 2019; 31: 71 – 80.

[57] Williamson G, Karazsia BT. The effect of functionality-focused and appearance-focused images of models

of mixed body sizes on women's state-oriented body appreciation. Body Image. 2018; 24: 95 – 101.

[58] Slater A, Cole N, Fardouly J. The effect of exposure to parodies of thin-ideal images on young women's body image and mood. Body Image. 2019; 29: 82 – 89.

[59] Barber C. [Internet] Available from: https://www.facebook.com/LADbible/posts/7767019503345172 [Accessed 2020-11-10].

[60] Dittmar H, Halliwell E, Ive S. Does Barbie make girls want to be thin? The effect of experimental exposure to images of dolls on the body image of 5- to 8- year- old girls. Developmental Psychology. 2006; 42(2): 283 – 292.

[61] Dove Evolution [Internet] Available from: https://www.youtube.com/watch?v=iYhCn0jf46U [Accessed 2020-11-10]

[62] Quigg SL, Want SC. Highlighting media modifications: Can a television commercial mitigate the effects of music videos on female appearance satisfaction? Body Image. 2011; 8: 135 – 142.

[63] Cohen R, Newton-John T, Slater A. The relationship between *Facebook* and *Instagram* appearance-focused activities and body image concerns in young women. Body Image. 2017; 23: 183 – 187.

[64] Cohen R, Irwin L, Newton-John T, Slater A. #bodypositivity: A content analysis of body positive accounts on Instagram. Body Image. 2019; 29: 47 – 57.

[65] Burnette CB, Kwitowski MA, Mazzeo SE. "I don't need people to tell me I'm pretty on social media:" A qualitative study of social media and body image in early adolescent girls. Body Image. 2017; 23: 114 – 125.

[66] Levine MP (February 1st 2017). Communication Challenges Within Eating Disorders: What People Say and What Individuals Hear, Eating Disorders - A Paradigm of the Biopsychosocial Model of Illness, Ignacio Jauregui-Lobera, IntechOpen, DOI: 10.5772/65305. Available from: https://www.intechopen.com/books/eating-disorders-a-paradigm-of-the-biopsychosocial-model-of-illness/communication-challenges-within-eating-disorders-what-people-say-and-what-individuals-hear

[67] Rodgers RF, Franko DL, Donovan E, Cousineau T, Yates K, McGowan K, Cook E, Lowy AS. Body image in emerging adults: The protective role of self-compassion. Body Image. 2017; 22: 148 – 155.

[68] Kelly AC, Miller KE, Stephen E. The benefits of being self-compassionate on days when interactions with body-focused others are frequent. Body Image. 2016; 19: 195 – 203.

[69] Golan M, Hagay N, Tamir S. The effect of "In Favor of Myself": Preventive program to enhance positive self and body image among adolescents. PLoS ONE. 2013; 8(11): e78223. doi:10.1371/journal.pone.0078223

[70] Alleva JM, Martijn C, Van Breukelen GJP, Jansen A, Karos K. Expand Your Horizon: A programme that improves body image and reduces self-objectification by training women to focus on body functionality. Body Image. 2015; 15: 81 – 89.

[71] Ziser K, Mölbert SC, Stuber F, Giel KE, Zipfel S, Junne F. Effectiveness of body image directed interventions in patients with anorexia nervosa: A systematic review. Int J Eat Disord. 2018;51:1121-1127. https://doi.org/10.1002/eat.22946

[72] Gattario KH, Frisén A. From negative to positive body image: Men's and women's journeys from early adolescence to emerging adulthood. Body Image. 2019; 28: 53 – 65.